Delivering xDSL

P9-EDF-138

Delivering xDSL

Lawrence Harte and Roman Kikta

McGraw-Hill

New York   San Francisco   Washington, D.C.   Auckland   Bogotá
Caracas   Lisbon   London   Madrid   Mexico City   Milan   Montreal
New Delhi   San Juan   Singapore   Sydney   Tokyo   Toronto

Cataloging-in-Publication Data is on file with the Library of Congress

1 2 3 4 5 6 7 8 9 0   DOC/DOC   0 9 8 7 6 5 4 3 2 1 0

ISBN 0-07-134837-9

*The sponsoring editor for this book was Steve Chapman, the editing supervisor
was Steven Melvin, and the production supervisor was Sherri Souffrance. It was
set in Times Roman.*

R. R. Donnelley & Sons Company was printer and binder.

*McGraw-Hill books are available at special quantity discounts to use as
premiums and sales promotions, or for use in corporate training programs. For
more information, please write to the Director of Special Sales, McGraw-Hill,
2 Penn Plaza, New York, NY 10121-2298. Or contact you local bookstore.*

This book is printed on recycled, acid-free paper containing a
minimum of 50% recycled, de-inked fiber.

Dedication

"I dedicate this book to my love Tara and my children Lawrence William and Danielle Elizabeth. To Tara, I love you to infinity and beyond."
    Lawrence

"I dedicate this book to: "Robert J. Charles, a telecommunications industry veteran who's bringing xDSL broadband access services to the United Kingdom, for his perseverance, dedication and inspiration."
    Roman

About the Authors

Lawrence Harte is the president of APDG, a provider of expert information to the telecommunications market.

Mr. Harte has over 21 years of experience in the electronics industry including company leadership, product management, development, marketing, design, and testing of telecommunications (cellular), radar, and microwave systems. He has been issued patents relating to cellular technology. He has authored over 75 articles on related subjects and has been a speaker and panel moderator at industry trade events.

Mr. Harte earned executive MBA at Wake Forest University and received his Bachelors degree from University of the State of New York. During the TDMA digital cellular standard development process, Mr. Harte served as an editor and voting company representative for the Telecommunications Industries Association (TIA) TR45.3, digital cellular standards committee.

As of 2000, Mr. Harte had authored and co-authored over 12 books relating to telecommunications technology. He has served as a consultant and expert witness for leading companies including Ericsson, Siemens, VLSI, AMD, Casio, Samsung, Sony, ATT, Nokia, Hughes and many others.

Roman Kikta is the Director of Strategy & Business Creation, Nokia Ventures Organization, a unit of Nokia Corporation. He is a seasoned telecommunications industry veteran, innovator and visionary.

During his 17+ years in telecommunications, he has held product planning & development, marketing and market / business development positions with leading telecommunications companies, including Nokia Mobile Phones, Panasonic, GoldStar, and OKI Telecom. Mr. Kikta has influenced several cellular mobile, transportable and portable phone model designs, features and functionality. His efforts have resulted in the first cellular payphone, first cellular PBX adjunct "Business Link," voice recognition dialers as well as the first PCS product launch in the US.

Mr. Kikta possesses exceptional foresight and understanding of global market needs, both from anthropological and psychological aspects, as well as of the opportunities provided by existing and future broadband, Internet and related technologies and businesses.

Mr. Kikta, a graduate of Rutgers University in New Jersey, is a co-author of the books: *CDMA IS-95 for Cellular and PCS: Technology, Economics & Services*; *Delivering xDSL*; *3G Cellular and PCS Demystified*; and *WAP Demystified*. He has also written several articles published in Industry magazines. Mr. Kikta has been a speaker at several telecommunications and Internet conferences and events, both in the US and Internationally.

Table of Contents

Foreword xvii

Preface xxi

Acknowledgments xxv

**CHAPTER 1.   INTRODUCTION TO TELEPHONE
               TECHNOLOGY** 1

ANALOG LOCAL LOOP 2
   End User Equipment 2
   Telephone Lines 3
   Switching 6
   Switch Interconnections 9
   Switch Control 10
   Telephone Companies 11
THE NEED FOR xDSL HIGH SPEED DATA CONNECTIONS 12
   Voice System Overload 13
   Data Transmission 14
   Digital Video Services 15
   High Bandwidth Competition 15
   Rapid and Cost Effective Roll-out 16

CHAPTER 2.   DIGITAL TRANSMISSION 17

DIGITAL LOCAL LOOP 18
   Integrated Services Digital Network (ISDN) 18
   Integrated Digital Loop Carrier (IDLC) 20
   Optical Systems 21
DIGITAL TRANSMISSION TECHNOLOGY 22
   Shared Analog and Digital Transmission 22
   Line Coding 24

Error Detection and Correction 27
Protocol Adaptation 30
TRANSMISSION MEDIUM LIMITATIONS 31
Frequency Response 31
Crosstalk (Signal Leakage) 33
Signal Ingress 34
Bridge Tap Reflections 35
Audio Loading Coils 36
Line Splice Attenuation 36
Line Resistance Attenuation 38

CHAPTER 3.   INTRODUCTION TO xDSL 39

HISTORY OF xDSL 40
xDSL TECHNOLOGIES 42
ISDN Digital Subscriber Line (IDSL) .43
High Bit Rate Digital Subscriber Line (HDSL) 43
Symmetric Digital Subscriber Line (SDSL) 45
High Bit Rate Digital Subscriber Line - 2nd Generation (HDSL2) 45
Asymmetric Digital Subscriber Line (ADSL) 46
Rate Adaptive Digital Subscriber Line (RADSL) 49
Consumer Digital Subscriber Line (CDSL) 49
Very High Bit Rate Digital Subscriber Line (VDSL) 50
TECHNOLOGY COMPARISON SUMMARY 53

CHAPTER 4.   OTHER BROADBAND TECHNOLOGIES 55

CABLE TELEVISION SYSTEM 56
First Generation (One-Way) Cable Modems 57
Hybrid Fiber Coax (HFC) 59
WIRELESS CABLE 61
Multmegabit Multimedia Distribution System (MMDS) 63
Local Multipoint Distribution Service (LMDS) 64
WIRELESS LOCAL LOOP (WLL) 65
Point-to-Multipoint (PMP) Microwave .69

WIRELESS LOCAL AREA NETWORK (WLAN) 70
MOBILE WIRELESS SYSTEMS 72
  Cellular and PCS Systems 72
SATELLITE SYSTEMS 75
FIBER SYSTEMS 79

CHAPTER 5.   END USER DSL EQUIPMENT **83**

xDSL MODEMS 84
  Ethernet Adapter 85
  External USB Modem 86
  Computer Modem (PCI) Card 87
  PCMCIA Card 88
SPLITTER 89
SET TOP BOX 89
RESIDENTIAL GATEWAY 89
PREMISES DISTRIBUTION NETWORK (PDN) 90
  Ethernet 91
  ATM 25 91
  Phoneline Networking 91
  Universal Serial Bus (USB) 92
  FireWire 92
PROXY SERVER 94
DIGITAL SERVICE UNIT (DSU)/CHANNEL SERVICE UNIT (CSU) 95
INTEGRATED ACCESS DEVICE (IAD) 95
SOFTWARE 96
SECURITY 96
  Network Address Translation (NAT) 97
  Dynamic Host Configuration Protocol (DHCP) 98
  Firewall 99
  IPSec 100
  Security Tips 100

INSTALLING END USER DSL EQUIPMENT 101
   Equipment Configuration Options 101
   Line Installation 103

CHAPTER 6.   DSL NETWORKS **107**

NETWORK OVERVIEW 107
   DSL Modem - Remote 109
   xDSL Splitter 109
   Network Termination 110
   Transmission Line 111
   Copper Cross Connect 111
   DSL Modem – Central Office 112
   Digital Subscriber Line Access Module (DSLAM) 113
NETWORK ROUTING AND SWITCHING EQUIPMENT 114
   Router 114
   ATM Switch 115
NETWORK MANAGEMENT 116

CHAPTER 7.   xDSL ECONOMICS **117**

END USER EQUIPMENT COST 121
   Development Costs 121
   Cost of Production 123
   Patent Royalty Cost 125
   Sales and Marketing Cost 125
   Post Sales Support 127
   Manufacturers' Profit 127
SYSTEM EQUIPMENT COST 127
   Development Costs 128
   Cost of Production 129
   Patent Royalty Cost 132
   Marketing Costs 133
   Post-Sales Support 134
   Manufacturer Profit 134

NETWORK CAPITAL COST 134
    Access Node/DSLAM 135
    System Operations Center 137
    DSL System Cost Case Study 137
OPERATIONS, ADMINISTRATION AND MAINTENANCE COST 139
    Leasing and Maintaining Interconnection Lines 139
    Interconnection Tariffs 141
    Billing Services 142
    Staffing and Maintenance of Access Lines 143
SERVICE REVENUE POTENTIAL 144
    Voice Service Cost to the Consumer 144
    Data Service Cost to the Consumer 145
    Internet Access 145
    Other Services 145
MARKETING CONSIDERATIONS 146
    End User Equipment Cost 146
    Interoperability Challenges 146
    Availability of Equipment 147
    Capacity Limitations 147
    Churn 147
    Distribution and Retail Channels .148
RESELLER COMPETITIVE TACTICS 149
    Low Cost Leased Line Conversion 149
    Internet Telephony 149
LOCAL CARRIER COMPETITIVE TACTICS 150
    Lack of Line Availability 150
    Installation of Filters 150

CHAPTER 8.   FUTURE HIGH SPEED DIGITAL TRANSMISSION
                TECHNOLOGIES 153

xDSL TRANSMISSION IMPROVEMENTS 153
    Advanced Echo Canceling 153
    Improved Line Coding 156
    Increased Bandwidth 156

Improved Cabling 158
POWERLINE DSL 158
Consumer Electronics Bus (CEBus) 161
HOME TELEPHONE LINE DISTRIBUTION 161

CHAPTER 9. DSL SERVICES **163**

VOICE OVER DSL (VoDSL) 164
Computer-to-Computer 167
Computer-to-Telephone 167
Telephone-to-Telephone 168
VIDEO OVER DSL 168
WebCam 172
Video Mail (VMail) 172
T1/E1 SERVICE OVER DSL 173
VIRTUAL PRIVATE NETWORKS (VPN) OVER DSL 173
POINT-TO-POINT PROTOCOL (PPP) OVER DSL 175
ATM OVER DSL 175

CHAPTER 10. BROADBAND APPLICATIONS **179**

DISTANCE LEARNING 182
Elementary (K-12) Education 184
Higher Education 184
Professional 186
Government 186
Military 187
ONLINE RETAIL 187
Travel 189
Software 189
Music 190

Books 190
Apparel 191
Food and Beverage 191
Consumer Electronics 191
Computer Hardware 192
ONLINE COMMERCE 192
Online Trading 193
Online Banking 193
Bill Presentation and Payment 194
ONLINE ENTERTAINMENT 195
Gambling 195
Networked Games 196
Interactive Toys 197
Movie Rental 197
Music Content 198
Interactive Movies 198
Virtual Radio Stations 199
Virtual Television Stations 199
Virtual Books/ E-Books 200
Virtual Newspapers and Magazines 201
Electronic Photo Album 202
BUSINESS APPLICATIONS 203
Video Conferencing 203
Remote Corporate Network Connections 203
Business Kiosks 204
Documentation Management .205
Field Service 206
Customer Care 207
MEDIA PRODUCTION 207
Image and Video Production .207
Printing Press 207
TELEMEDICINE 208
Remote Clinical Facilities 209
MANUFACTURING 210
Production Monitoring 210

ADVANCED COMMUNICATIONS SERVICES 211
   Telephone Network Bypass 211
   Internet Telephony 212
   Electronic Mail (email) 212
   High Speed Network Interconnection 213

APPENDIX I.   ACRONYMS AND ABBREVIATIONS 217

APPENDIX II.   GLOSSARY 225

APPENDIX III.   UNITED STATES xDSL PROVIDERS 271

INDEX 275

Foreword

Many Americans have no idea what DSL means or what it will bring, yet DSL promises profound changes in information exchange and communications. Our company, IP Communications, describes these exciting new possibilities as the "e-experience" – rich, diverse, accessible and convenient. Video. Multimedia. Music. Telephony. With DSL, the medium **enables,** rather than limits, the message.

Of the emerging broadband technologies, the fastest growing is DSL. While cable provides quick, "always on" connections, it is not available to most businesses and speeds drop dramatically as neighborhood traffic increases. Satellite is available nearly everywhere, but it is expensive and has limitations (the service must be split between the satellite for downloads and modem connections for uploads). Fixed wireless has limited availability and carries a high price.

DSL will be the leading technology, because it provides businesses and consumers what they need over the **current** network infrastructure. Using existing copper wires, DSL typically provides connection speeds from 384kbps to 6Mbps, with pricing that makes it affordable to most users. The connection is always on, the installation and turn-up are relatively simple, and the service reliability is high.

Today, 95 percent of consumers and businesses use old-fashioned dial-up Internet access service, at speeds of 56k or less. Think Model T. Within the next few years, it's estimated the majority of these users will have moved to DSL service at speeds 50-100 times faster than they have been used to. Think Ferrari.

Speed really isn't the most important part of the DSL story, however. Techies focus on the characteristics of DSL, but consumers and businesses are interested in *applications*. I am reminded of Mark Twain's comment upon hearing of plans to build the first transcontinental telegraph line. "That's all well and good," he said, "but what will people *say*?" Twain understood that the technology is only as good as the applications it enables.

That's why I was pleased to be asked to write the foreword for *Delivering xDSL*. The authors do a superb job of describing the applications, technology and engineering behind DSL. What makes this book especially relevant to all of us – engineers, marketers, administrators, as well as consumers – is its explanation of the possible and the probable.

One of my marketing managers describes the changes ahead as "moving the power of the Internet into the kitchen." His point is that the kitchen is the place people tend to perform many routine tasks – finding a phone number, paying bills, keeping track of schedules, making phone calls, preparing meals. Many of the applications that help people do these things quickly and easily online already exist. People don't use them because it's inconvenient to *wait* for dial-up access, *wait* for applications to load, and *wait* for information to download. When people don't have to wait, they will start using the Internet for everyday tasks; and that, he argues, is when the market for DSL and broadband services will take off.

First, however, the industry needs to finish building a solid foundation. I'm concerned that the market segmentation strategies of most companies are helping create a "digital divide." That's why IP Communications is "saturating" the states we are entering. We are serving metropolitan *and* rural areas. In some ways, DSL is more important to the rural areas and small cities because it levels the playing field, providing big-city capabilities for business and consumer. As this market evolves, expect market segmentation to lead to consolidation and ubiquity.

W. Dal Berry
Chairman & CEO
IP Communications, Inc.
Dallas, Texas
September 2, 2000

Preface

At the beginning of the 21st century, telephone technology had changed so significantly over the previous 5 years, the jobs of almost all professionals in the telephone industry had been impacted by a specific new technology. This new technology allows a standard twisted pair of telephone wires to provide high quality television service, audio broadcast, high speed internet access and many other information services direct to consumers in their homes. In the year 2000, there were over 700 million wired telephone customers throughout the world that are connected by a twisted pair of wires. Almost all of these customers can benefit from the new digital subscriber line technology.

To further complicate the lives of telephone professionals, several competing digital subscriber line (DSL) technologies have developed with the proponents (developers) of each sometimes offering conflicting information. While each of these technologies allow the conversion of these twisted wires into high speed digital lines, they each have advantages and limitations when compared with each other. This book provides an unbiased source of information on DSL technologies.

Digital subscriber lines that are being introduced to the market offer advanced features, services, and cost advantages over the older analog telephone technologies. This book is a guide which provides a big picture of these digital subscriber line technologies, features, costs, and services that make them very desirable to have.

This book offers a balance between marketing, applications, services and technical issues. It covers what's new in digital subscriber line technology, explains how it works using over 200 illustrations, and describes the marketing aspects of the new technologies and services. Over 100 industry experts have reviewed the technical content of this book. Many of the industry buzzwords are defined and explained.

To help meet the growing demand for cost effective DSL service and advanced features, several DSL technologies are available or are in testing stages in over 60% of the world telephone systems. Theoretically, analog telephone systems might have provided for advanced services, but for a variety of reasons, they have not. Furthermore, each DSL technology's unique advantages and limitations offer important economic and technical choices for managers, engineers, and others involved with telephone systems. Delivering xDSL provides the background for a good understanding of the technologies, issues, and options available.

The chapters in this book are organized to help technical and non-technical readers alike to find the information they need. Chapters are divided to cover specific technologies, economics, and services. The chapters may be read either consecutively or individually.

Chapter 1: Provides an introduction to telephone technology. The technologies throughout the book evolved from analog telephone systems. A description of basic analog and digital telephony is included. This chapter is an excellent introduction for newcomers to telephone technology.

Chapter 2: Describes digital telephone technology and services. It includes a semi-technical description of advanced digital services common to all digital technologies, summarizing potential new services and how they may be implemented.

Chapter 3: Explains the different types of xDSL technologies. This includes IDSL, SDSL, HDSL, HDSL2, ADSL, ADSL-Lite, VDSL and variants of these technologies. A brief history of the key functions for each technology and why they were developed is provided here.

Chapter 4: Identifies and describes competing high bandwidth access technologies. This chapter covers the basics of high bandwidth technologies. This includes fixed wireless, fiberoptic systems, cable modems, and satellite systems. A basic overview of each technology is provided along with descriptions of their strengths and limitations.

Chapter 5: Provides the fundamentals of DSL end user equipment. This includes video and computer network interconnection equipment, telephone devices and differences between xDSL equipment options.

Chapter 6: Includes DSL network requirements, equipment, implementation methods, a high-level overview of the public switched telephone network (PSTN), network interconnections, switching equipment, and system planning.

Chapter 7: Reviews xDSL system economics, including costs of digital subscriber units and system equipment, and an analysis of telephone network capital and operational costs. Revenue producing services, distribution channels, and marketing costs are explained.

Chapter 8: Describes future DSL technologies. This chapter covers the latest advances and the predicted evolution of xDSL technologies.

Chapter 9: Provides an overview of broadband services. Covers video services, Internet access, high bandwidth bypass, virtual private networks (VPN) and asynchronous transfer mode (ATM).

Chapter 10: Covers key applications that benefit from DSL. These applications require the speed of DSL. This chapter provides current and projected market demand for these applications. Described are distance education, movie rentals, virtual radio stations, distance learning, shopping services with images, interactive gaming and many others.

Appendices are provided which include acronyms, definitions, sources of industry specifications and a listing of xDSL providers in the United States.

Acknowledgments

We thank the many gifted people who gave their technical and emotional support for the creation of this book. In many cases, published sources were not available on this subject area. Experts from manufacturers, service providers, trade associations and other telecommunications related companies gave their personal precious time to help us and for this we sincerely thank and respect them.

We especially thank Gloria Consola from 2Wire, Mark Burn with Alcatel, Steve Wilhite at Apple Computer Corporation, Emily Padlan of Cisco Systems, Inc., Connie Henry from Copper Mountain Networks, Inc., Lori Hicks with Efficient Networks, Dave Morris at Innovative Connectors ("InCon"), W. Dal Berry with IP Communications, Sue Cifelli from mPhase Technologies, Sarah Hopps of Net2Phone, Helen Choy with Netopia, Inc., Kristine Freitas with SonicWALL, Terri Tiffee at SourceCOM, Natalie Banozitz of Xpeed Networks, Inc., and Robert J. Charles with SpeedLink.

Special thanks to the people who assisted with the production of this book including: Amy Case (project manager), Mary Case (illustrator), Lawrence William Harte (researcher and illustrator), Danielle Elizabeth Harte (illustrator), and Darian Black (layout). Special thanks to Steve Chapman at McGraw-Hill who helped ensure this book was at the highest industry standard and that the book contained valuable and quality information. And thanks to our financial supporters including Linda Plano, Konny Zsigo, Mike Cromie, Ted Ericsson, Eric Stasik, Micheal Zapata, Elliott Hamilton, Quincy Scott, and Virginia Harte

Chapter 1

Introduction to Telephone Technology

Telephone technology involves the transfer of information from one communication device to one or more other communication devices. The evolution of communications technology in the late 1990's has allowed the term information to include voice, data and video services. What telecommunications technology does not guarantee is the cost effective delivery of these services nor does it guarantee the availability of applications that are useful through interconnection to customers.

Since the 1870's, telephone technology has primarily been focused on transferring audio (analog) signals between two points. Early telephones were directly connected to each other. As many more telephones were produced than lines, switching systems were developed to allow the interconnection of any individual telephone. Initially, these switching systems were manually connected patch panels. Eventually, the switching systems became automated to allow direct dialing of phone calls.

In the 1990's, the Internet and data communication systems began to require a new type of communication system. These systems required the transfer of high-speed data signals to a variety of data networks. The switching systems that are used for voice switching are not appropriate for high-speed data communications. These include video services and data network connections.

Analog Local Loop

In the year 2000, there were over 700 million analog telephone lines throughout the world. A majority of telephone lines use pairs of copper wire to carry audio signals. Although these copper wires were designed to carry voice signals, they are capable of carrying low speed data signals through the use of audio modems. Audio modems convert digital signals to audio tones so they can be transferred through the existing telephone network.

End User Equipment

Most of the telephone equipment in use during the year 2000 converted electrical analog (audio signals) into acoustic energy that the customer can hear. The basic function of analog telephone service is called plain old telephone service (POTS).

Figure 1.1 shows a block diagram of a standard POTS telephone (also known as a 2500 series phone). This telephone continuously monitors the voltage on the telephone line to determine if an incoming ring signal (high voltage tone) is present. When the ring signal is received, the telephone alerts the user through an audio tone (on the ringer). After the customer has picked up the phone, the hook switch is connected. This reduces the line connection resistance (through the hybrid) and this results in a drop in line voltage (typically from 48 VDC to a few volts). This change in voltage is sensed by the telephone switching system and the call is connected. When the customer hangs up the phone, the hook switch is opened increasing the resistance to the line connection. This results in an increase in the line voltage. The increased line voltage is then sensed by the telephone switching system and the call is disconnected.

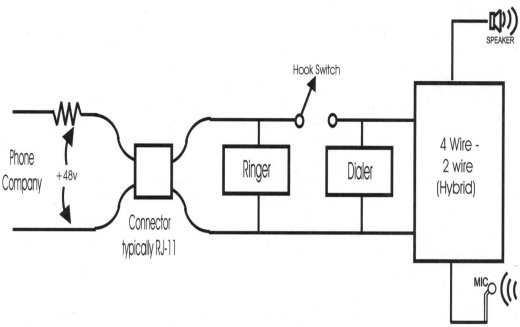

Figure 1.1, POTS Telephone Block Diagram

Telephone Lines

Since the early 1900's, copper telephone lines have been the primary method of connecting customers to the telephone system. These telephone lines typically use twisted pairs of copper wire. The twisting of wires reduces the effects of electrical noise from distorting the desired audio signals. In essence, when the noise is received on one twist, the same noise is received on the other twist. The voltage goes positive on one line while it also goes positive on the other. Basically, the two signals are at the same level and they cancel each other (balance).

Telephone lines usually start from the central office's switching center in the form of bundles of many wire pairs (trunks). These trunks connect the central switching office to distribution cables (cables with a reduced number of wire pairs) that eventually are connected to individual homes or businesses. Trunks may contain thousands of pairs of wires while local distribution cables only contain 25 to 100.

Cables are produced in rolls with a limited length (often 500 feet long). The installation of telephone cables requires several splices points as the large trunk cables connect to the distribution cables that connect to the drop cables to the home.

Figure 1.2 shows a typical layout of a telephone wiring between the central office and the home. This diagram shows that 600 pairs of copper line start from the central office (CO). This trunk cable is connected to (3) 200 pair distribution cables that supply circuits to nearby neighborhoods. As the cables enter into a neighborhood, they are connected at splice points to smaller distribution cables until a final distribution cable that only holds 25 pairs reaches a telephone pole

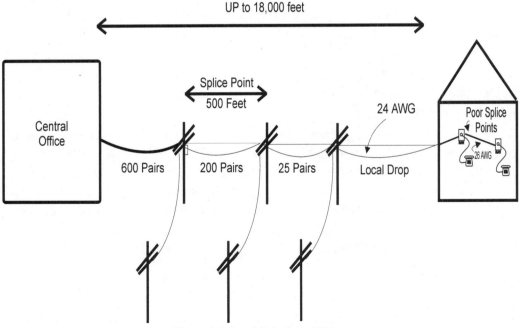

Figure 1.2, Local Telephone Wiring

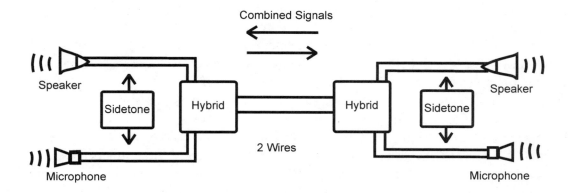

Figure 1.3, Analog Telephone Transmission

located near a house. At the telephone pole, usually 2 pairs of wires are tapped to the drop line that enters into the house (to allow up to 2 separate phone lines). These 2 pairs of wire are attached to a network interface device (NID) that protects (isolates) the wiring in the home from the telephone network wiring. Once in the home, twisted pairs of wires are looped from the NID to telephone jacks within the house. This illustration also shows that there is significant potential for different types and sizes of wire and many splice points. This inconsistency can dramatically affect the ability to transfer high-speed digital signals.

Until the 1990's, telephone transmission had remained basically the same since the late 1800's. Acoustic energy from the customer is converted to electrical signal by a microphone in a handset. This electrical energy was applied through a hybrid electrical device to the telephone line through the speaker in the handset. A telephone hybrid device (often called a "magic" device by telephone personnel) allows energy from the microphone to enter into the power line while extracting most of the microphone energy from the speaker. At the same time, microphone energy that is received from the distant handset is applied to the speaker but not to the microphone.

Figure 1.3 shows how a typical analog telephone transmission line operates. In this diagram, audio from customer #1 is converted to electrical energy by microphone #1. This signal is applied to the telephone line via the hybrid adapter #1. A portion of this signal is applied to the handset speaker to produce sidetone (so the customer can faintly hear what they are saying). This audio signal travels down the telephone line to hybrid #2. Hybrid #2 applies this signal to speaker #2 so customer #2 can hear the audio from customer #1. When customer #2 begins to speak, microphone number #2 converts the audio to an electrical signal. This signal travels down the line to hybrid #1. Hybrid #1 subtracts the energy from microphone #1 (the combination of both signals are actually on the line) and applies the difference (audio from customer #2) to the speaker #1.

Switching

The function of a switching system is to connect two (or more) points together. These connections can be physically connected (mechanical switch) or connected logically (through software). The first telephone systems performed the switching of calls by human operators. The operators interconnected telephone lines by manually connecting cables at switchboards. These switchboards contained many wires that had plugs and the switchboard had many sockets for the plugs. To interconnect telephone calls at long distances, one operator would have to call other operators to setup the call. Setting up calls could be a complex process and this process got more complex as many more telephones were installed.

Switching systems have evolved from manual switchboard systems (wires and plugs) to logical (digital) switches. The earlier types of manual switchboard systems were changed to automatic switching systems to eliminate the need for operators to setup every call. The first types of automatic switching systems used crossbar switches. Crossbar switches used mechanical arms to physically connect wires (or busses) together.

Figure 1.4 shows a crossbar switching system. In this example, there is a matrix of lines (busses) where each input line can be connected to any output line. When a connection needs to be made, a mechanical switch connects one of the busses with the other busses. The disadvantage of this system is that the number of mechanical switches for connecting each input port to an output port exponentially increases with the number of ports that require connection. For example, a switch with 10 inputs and 10 output lines requires 100 switches. A switch that has 20 inputs and 20 outputs requires 400 switches.

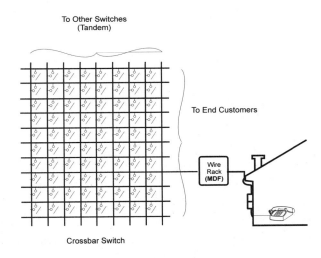

Figure 1.4, Crossbar Switching

In the late 1960's, mechanical switching systems began to change to time slot interchange (TSI) digital switching systems. Time slot interchange switches use memory banks and software control to logically connect lines to each other.

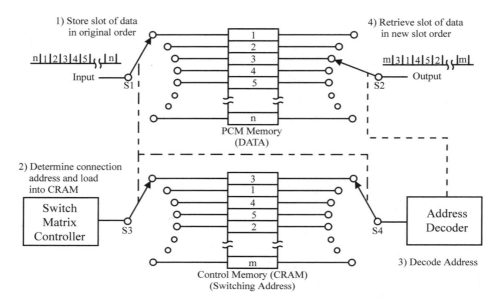

Note: Switches (S1-S4) are electronic and do not use mechanical parts.
Figure 1.5, Time Slot Interchange (TSI) Switching

Figure 1.5 shows a TSI switching system. This diagram shows a simplified matrix switching system. Each input line (port) is connected to a multiplexer. The multiplexer places data from each port in time sequence (time slot) on a communications line (e.g. a T1 or E1 line). This time multiplexed signal is supplied to a matrix switching assembly. The matrix switching assembly core has two memory parts; a section that holds the pulse coded modulation (PCM) data and Control Memory - CRAM that holds switching addresses data.

The time slots (voice channels) from the incoming multiplexed sent through switch S1 to be sequentially stored in the PCM data memory. The data is later retrieved by switch S2 and placed on a specific time slot on an outgoing line. The outgoing multiplexed line is supplied to a de-multiplexer so each time slot is routed to an output port.

Switch Interconnections

The public telephone networks involve the interconnection of telephone switches to each other. Local switches must be connected to other local switches, long distance switches and foreign exchanges. These switches are interconnected to each other through high-speed communications lines.

For business reasons, there are demarcation points in the network to allow the switches of different companies to interconnect to each other. These connection points are often called points of presence (POP).

Figure 1.6 shows a typical public switched telephone network (PSTN). Notice that geographic areas controlled by a local telephone company can have several switching systems (called a "Central Office (CO)"). These COs are usually direct connected with each other. When communications must progress out of the geographic area, the signal is routed through the POP to another carrier, typically an Inter-eXchange Carrier (IXC) or Foreign EXchange (FEX).

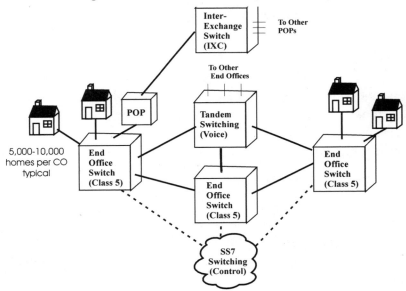

Figure 1.6, Public Switched Telephone Network (PSTN)

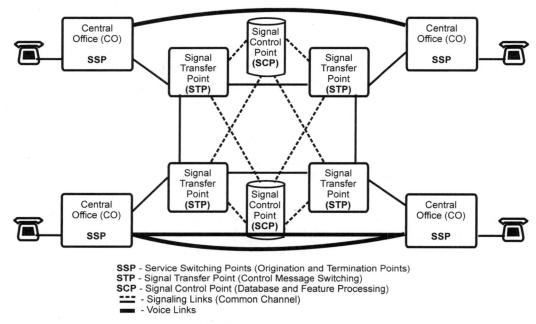

SSP - Service Switching Points (Origination and Termination Points)
STP - Signal Transfer Point (Control Message Switching)
SCP - Signal Control Point (Database and Feature Processing)
‑‑‑ - Signaling Links (Common Channel)
▬ - Voice Links

Figure 1.7, SS7 Common Channel Signaling

Switch Control

In the early telephone systems, the control of a telephone call was coordinated by control signals or tones that were sent on the same audio lines as the voice signals. These control tones would either be mixed with the audio or temporarily replace the audio signals. This audio signal control is called in-band signaling.

As the design of telephone systems advanced, it was necessary to add more intelligence to the call setup (e.g. automatic forwarding of telephone calls). It became necessary to shift the control signaling to circuits outside the audio path. This allowed more rapid call setup and better overall control over the communications connection.

Common channel signaling system #7 (commonly called "SS7") is the primary system used for interconnection of telephone systems. SS7 sends packets of control information between switching systems. Figure 1.7 shows the basics of the SS7 control signaling system. The SS7 network is composed of its own data packet switches, and these switching facilities are called signal transfer points (STPs). In some cases, when advanced intelligent network services are provided, STPs may communicate with signal control points (SCPs) to process advanced telephone services. STPs are the telephone network switching point that route control messages to other switching points. Signaling Control Points (SCP) are databases that allow messages to be processed as they pass through the network (such as calling card information or call forwarding information).

Telephone Companies

In the United States, telephone companies are regulated by the government, but not owned by the government. For most European countries and many other countries, local telephone service is provided by government owned posts, telephone and telegraph (PTT) operators. In some European countries, the post (mail) network has been separated from the operation of telephone and telegraph networks. In some countries, the telephone and telegraph systems have become privatized, and are no longer owned by the government.

To connect local systems to each other (long distance), inter-exchange service is provided. In the US, these are called inter-exchange carriers (IXCs). In the US, from 1984 until 1997, IXC and LEC operating companies were legally required to refrain from engaging in directly competitive business operations with each other. Since 1997, one business entity can engage in both IXC and LEC business if it satisfies certain competitive legal rules. In Europe and throughout the rest of the world, the same PTT operators also usually provide inter-exchange service within their country. In any case, governments regulate how networks are allowed to interconnect to local and long distance networks.

For inter-exchange connection, networks as a rule connect to long distance networks through a separate toll center (tandem switch). In the United States, this toll center is called a point of presence (POP) connection.

Networks often connect to other networks (such as the Internet or corporate networks) via a gateway connection. Various types of gateway connections can connect the Central Office (CO) to other public (e.g. Internet) or private (e.g. corporate) networks. A gateway transforms data that is received from one network into a format that can be used by a different network. A gateway usually has more intelligence (processing function) than a bridge as it can adjust the protocols and timing between two dissimilar computer systems or data networks. A gateway can also be a router when its key function is to switch data between network points.

Interconnections to other networks are classified by type of connection. Basically, the lower the connection type number, the simpler (and more limited) is the connection. The connection types include the basic customer type POTS (type 1) and inter-switch types 2. Type 1 POTS connection provides for basic signaling and low speed (audio) connection. The higher types of connection include various capabilities such as types of information services available (operator assist, emergency number support). In the United States, the typical interconnection types include those designated as type 2A, 2B and other variants of type 2, each serving a specific purpose. Type 2 interconnection connections link the LEC into a tandem (standard local switch interconnect) office. When using the type 2 connections, the CO appears as a standard end office switching facility.

The Need for xDSL High Speed Data Connections

The 1990's brought new key issues to the incumbent telephone companies; overburdening of switching systems due to the growth of the Internet, the demand for high bandwidth services without the installation of new lines and reduced willingness to invest in new technologies.

New digital transmission technologies were developed to allow the local loop to receive high-speed data signals. There are now several forms of digital subscriber lines. Each of these DSL technologies usually has a prefix to indicate the variant of DSL technology. Hence, the "x" in xDSL indicates that there are many forms of xDSL technology.

The key driving needs for xDSL systems include reducing the overload of voice switching systems, permit high-speed data transfer and provide for advanced information services such as digital video and to compete against new companies that are entering into the telecommunications marketplace.

Voice System Overload

The growth in the voice marketplace is small compared to the growth of data transmission through the Internet. Voice switching systems are designed to permit customers to originate and receive short telephone calls. Until the 1990's, customers used their telephones for approximately 30 minutes per day and rarely at the same time. This allowed the telephone service providers to service many customers with switching systems that could only complete calls for a small portion of customers.

When the Internet became popular in the late 1990's, the average telephone line usage for customers increased dramatically. In some cases, this caused the local voice switching systems to become blocked. The ability to separate data signals (such as data that is used to access the Internet) from the voice signals that go to the telephone network voice switching systems allows for a reduction in voice system overload.

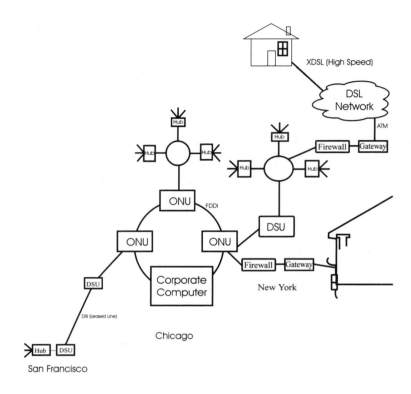

Figure 1.8, Wide Area Data Network

Data Transmission

High-speed data transmission service allows for rapid file transfer from networks such as the Internet. The need for cost effective long distance corporate computer network connections (wide area networks) has dramatically increased the number of high-speed alternatives.

Figure 1.8 shows how high speed data transmission services allow a corporation to extend its high speed corporate network to remote offices and to teleworkers (small offices and home based workers). This figure shows that one (or more) corporate communication networks can be interconnected using fiber distributed digital interface (FDDI). The FDDI provides up to 1 Gbps of data transfer.

Although 1 Gbps is being transferred in the network, each user only requires a few Mbps to transfer large files or digital images to and from the network. The corporate network contains a gateway that adapts the corporate network connection to a high-speed data transmission connection. Prior to allowing external connections, a "firewall" security device ensures only authorized computers can connect to the corporate network. At the other end of the high-speed data line connection, a satellite corporate office also has a "firewall" security device and a gateway. This gateway adapts the high-speed data connection to its 100 BaseT (standard 100 Mbps transmission system) computer network.

Digital Video Services

Digital video services include distance learning (education), movies (entertainment), telemedicine (professional services), security monitoring (safety), and other image-based types of services. Digital video signals require approximately 1.5 Mbps for a reasonable quality digital picture and digital high definition television (HDTV) signals require over 10 Mbps.

In the late 1990's, the Internet and television began to merge. Although the initial Internet television video interfaces were limited to web browsers, these devices added value to video broadcast services through advanced directory services.

High Bandwidth Competition

Companies and consumers are demanding higher data transmission bandwidths to allow for digital video and rapid data file transfer. In the 1990s, several new high bandwidth data transmission services became available through a variety of changes to existing transmission systems. These include cable television systems (cable modems), satellite systems (computer set top boxes), microwave cable television (MMDS and LMDS). High bandwidth services are now provided by incumbent local exchange companies (ILECs) and competitive local exchange companies (CLECs).

Companies that offer high bandwidth data transmission services can easily dedicate a portion of the data services for voice services. Although this requires the competing company to fall into a new regulatory category, the bundling of data and voice service is seen as a strong competitive advantage. ILECs are likely to aggressively compete against CLECs that offer alternative solution voice services by offering their own high-speed data and digital video services.

Rapid and Cost Effective Roll-out

Technologies continue to change and improve. xDSL technology allows ILECs and CLECs to gradually roll-out high speed data services without significant investment.

It is often hard to for a customer to decide which technology to select and implement due to the fear of choosing a technology that will become outdated and the recovery of investment may not be realized. The best example of this decision occurred when many local telephone service providers in the United States decided not to invest in Integrated Services Digital Network (ISDN) technology in the 1980's. ISDN technology allowed for much higher digital transmission rates on copper wires compared to the data transmission rates available from analog modems (144 kbps compared to 2.4 kbps or 9.6 kbps). To offer ISDN service, ILECs were required to pay for expensive upgrades to every switch in their network. If only a few customers took advantage of ISDN, the investment in new technology would be lost. In hind site, ILECs probably made the best choice as xDSL technology can be selectively added. The chosen technology can also be easily changed as new advances are made (e.g. upgrading from HDSL to ADSL to VDSL).

Chapter 2
Digital Transmission

Digital telephone transmission technology has evolved over the past 40 years to allow standard copper telephone wires (local loop) to carry up to 52 Mbps. The initial use of digital transmission was to allow a single line (or pairs of lines) to carry more than one communication channel. These systems are called pairgain systems as a single pair of copper wires can carry (gain) more channel carrying capacity.

Although the first uses of digital transmission was for interconnection trunk lines, digital transmission systems have been used in the local loop that include integrated services digital network (ISDN), integrated digital loop carrier (IDLC) and fiber optic distribution systems.

Digital communications systems utilize several recently developed technologies. These technologies include shared signal technology (multiple access and a single medium), advanced modulation, error correction and detection capability and protocol adaptation. Some DSL systems allow for the sharing of analog and digital signals on the same line. Highly efficient line coding (modulation and data structure) allows for high-speed data transmission on a copper wire pair.

Error correction and detection processes provide for reliable use of digital transmission lines. Protocol adaptation allows multiple types of systems to use digital transmission lines to interconnect these systems.

Digital transmission technology has evolved to overcome many of the physical limitations of copper wire signal transmission. These limitations include: poor frequency response, crosstalk (signal leakage), interference from other electric noise or radio signals, inconsistent line sizes, multiple connections, high frequency attenuation due to audio filters, attenuation at corroded lines splices, and the high resistance of long copper wireless to electrical signals.

Digital Local Loop

One of the first advances in local loop digital transmission technology over the local loop, was integrated services digital network (ISDN). ISDN systems were supposed to replace analog systems with high-speed digital channels. Although many ISDN systems were installed in Europe, the high cost of implementation and availability of alternative solutions delayed introduction into the United States. In 1999, over 15% of the distribution lines in neighborhoods used digital loop carrier (DLC) systems to serve homes cost effectively in the United States. Fiber to the home (FTTH) systems have been successfully tested, however the high cost of fiber cable installation limits the deployment of FTTH systems.

Integrated Services Digital Network (ISDN)

ISDN provides several communication channels to customers via local loop lines through a standardized digital transmission line. ISDN is provided in two interface formats; a basic rate (primarily for consumers) and high-speed rate (primarily for businesses). The Basic Rate Interface (BRI) is 144 kbps and is divided into three digital channels called 2B + D. The Primary Rate Interface (PRI) is 1.54 Mbps and is divided into 23B + D. The digital channels for the BRI are carried over a single, unshielded, twisted pair, copper wire and the PRI is normally carried on (2) twisted pairs of copper wire.

The "B" channels operate at 64kb per second digital synchronous rate and the "D" channel is a control channel. The D channel is used to coordinate (signal) the communication with the telephone network. When used on the BRI line, the D channel is 16kbps and when provided on the PRI channel, the D channel is 64 kbps. Because the amount of telephone system control signaling is relatively small, the D channel can also be used for low speed packet data messaging. The 64 kbps "B" channels can be used for voice, data. On the BRI system, the two B channels can be combined for 128 kbps data connection.

ISDN telephone lines exclusively use digital transmission. This requires a customer to replace their analog telephones with ISDN digital telephone equipment if they upgrade to ISDN service. ISDN service is typically provided using modular plugs. These plugs include a RJ45 interface (8 pin) for data equipment (called a BRI-S/T) and the other physical connection type is a two-wire, RJ11 type standard (called the BRI-U).

The maximum distance for a BRI-S/T line is approximately 3,000 feet and the maximum distance for the BRI-U is 18,000 feet. Beyond these distances, the service provider may install repeaters to provide service. However, repeaters are expensive to install and setup.

The ISDN BRI allows the user to change the use of the B channels whenever desired. For example, an ISDN user may be sending data using the two B channels at 128 Kbps. If a voice call comes in or is initiated, the data transmission is not interrupted; but is automatically reduced to one B channel at 64 Kbps. When the voice call ends, the data transmission returns to 128 Kbps on the two B channels.

Integrated Digital Loop Carrier (IDLC)

Integrated digital loop carrier (DLC) is a digital transmission technology that is used between the central office and groups of customers. The IDLC system is composed of two primary parts; an integrated digital terminal (IDT) and a remote digital terminal (RDT). The IDT concentrates up to 96 lines on to a single 24 channel T1 line. It does this by assigning central office channels to time slots on the IDLC line (between the IDT and RDT) as needed. The RDT reverses the process by assigning a time slot to an access line. The RDT also changes the format of the time slot to the access technology of choice (e.g. ISDN or analog).

The key advantages to DLC carrier systems are the cost effective transmission and the ability to rapidly add, delete or change customer services without having to dispatch an installation technician. The DLC system offers improved efficiency through the use of existing distribution cabling systems. DLC systems also offer the ability to extend the range of access lines from the central office to the end customer as the RDT effectively operates as a repeater.

Unfortunately, DLC systems are not transparent to xDSL systems. Although it is possible to install digital subscriber line network equipment (co-locate) along with RDT equipment, the RDT equipment housings and power supplies were not originally designed to hold additional equipment.

An RDT is divided into three major parts; digital transmission facility interface, common system interface and line interface. The digital transmission interface terminates the high-speed line and coordinates the signaling. The common system interface performs the multiplexing/de-multiplexing, signaling insertion and extraction. The line interface contains digital to analog conversions (if the access line is analog) or digital formatting (if the line is digital).

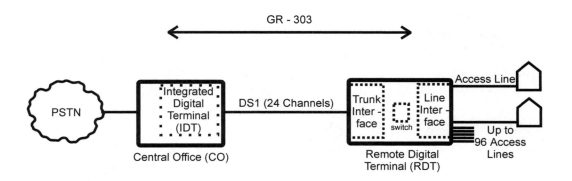

Figure 2.1, Digital Loop Carrier (DLC) System

Figure 2.1 shows an integrated digital loop carrier system. The DLC carrier system is composed of two basic parts; the Integrated Digital Terminal (IDT) and the Remote RDT. The IDT dynamically connects access lines (actually digital time slots) in the switching system to time slots on the communications line between the IDT and RDT. The RDT reverses the process and converts the high speed DLC channel into independent POTS channels (DS0s) that are connected via local loop lines to homes or businesses.

Optical Systems

Optical systems transfer information in optical form. The data transmission bandwidth of optical systems can exceed 10 Gbps. Optical fiber systems have a much longer life than copper cables as fibers are not subject to the corrosive effects of water and a single fiber can carry thousands of voice channels.

Some of the disadvantages of fiber when it is used in the local access loop include a higher cost of cable per meter, a higher cost for splicing and more costly electrical interfaces. Fiber cables cost approximately 20 cents per meter of fiber compared to 10 cents per meter of twisted copper pair wire [1]. The complexity of splicing and installing connectors to fiber cables is much higher than connecting copper wires together. This results in higher cost of fiber than copper when it is installed in a local loop. When fiber is used to carry multiple circuits (e.g. between switching systems), the effective cost of fiber per communications circuit is substantially lower than the cost of copper.

Digital Transmission Technology

The original reason for digital transmission technology was to increase the efficiency of copper wire to carry more communications channels. Ironically, because the cost of digital signal processing in the 1960s was high, the first digital transmission system was much more expensive than its analog counterparts.

Although digital transmission over copper wire involves the transfer of binary information, the signals that carry the digital information are analog (continuously changing) signals. These signals can experience and can cause distortion to nearby copper wires. Various technologies have been developed that allow for high speed data transmission on copper wires. These include the ability to share analog and digital signals on the same copper pair, more efficient transmission of digital information, hybrid transmission technology (simultaneous signal transmission) and standardized adaptation from one communications system to another (protocol adaptation).

Shared Analog and Digital Transmission

Because of the large number of installed analog telephones, it is very desirable to provide both analog and digital service without requiring the replacement of existing telephone equipment. Early digital transmission systems (e.g. ISDN and T1) digitized the entire bandwidth of the copper wire transmission lines. This required either the replacement of all existing analog telephones or the installation of an active line adapter to convert the digital. Through recent developments in technology, both analog (standard telephone) and high-speed digital signals can share the same copper telephone line. The sharing of these signals allows selective upgrading for existing customers without changing the installed equipment and with minimal changes of lines.

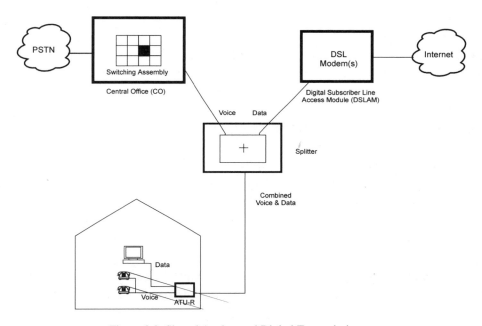

Figure 2.2, Shared Analog and Digital Transmission

Figure 2.2 shows that analog and digital signals can share the same transmission line by using 2 different frequency bands. The diagram shows that approximately 1-2 MHz of usable bandwidth can be used on a copper line. The analog (low frequency signal) occupies the lower portion of the frequency band (4 kHz and below). The audio signal can remain on the same while the digital (high speed data) modulates a RF carrier signal that has a frequency range located above the audio signal.

Line Coding

Line coding is the format of information that is transmitted on the line. Line coding for xDSL systems primarily encompasses modulation type. The type of line coding determines the efficiency of data transmission (bits per Hz of line bandwidth), the maximum transmission distance as specific transmission data rates and the ability of the transmission to tolerate distortions on the line.

The different types of modulation used in xDSL systems include Quadrature Amplitude Modulation (QAM), Carrierless Amplitude and Phase (CAP) and Discrete Multitone Transmission (DMT).

Quadrature Amplitude and Phase Modulation (QAM)

Quadrature Amplitude and phase Modulation (QAM) is an efficient modulation system that combines two different types of modulation (amplitude and phase) to deliver a large number of bits per available Hertz of bandwidth. Figure 2.3 shows the process of combining phase and amplitude modulation. In this example, one digital signal changes the phase and another digital signal changes the amplitude. In some commercial systems, a single digital signal is used to change both the phase and the amplitude of the RF signal. This allows a much higher data transfer rate as compared to a single modulation type.

Carrierless Amplitude and Phase (CAP) Modulation

Carrierless amplitude and phase (CAP) modulation is very similar to QAM modulation. The difference is the continuous shifting of phase (or signal mix) of the carrier signal level. CAP modulation was designed to help to reduce the effects of crosstalk and to simplify the signal processing of modulated signal. CAP transmits data signals on a single high bandwidth modulated carrier.

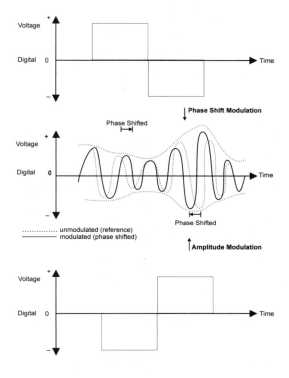

Figure 2.3, Quadrature Amplitude Modulation (QAM)

Because CAP modulation requires only a single downstream and upstream carrier, the modulation is simple compared to DMT (discussed next). This consumes less power and creates less heat as a result. While this may not seem very important, the amount of power consumption and heat that can be dissipated in small equipment cabinets is very limited.

Discrete Multitone (DMT)

A recent application of data transmission on copper wire involves the transmission of several narrow sub-channels instead of a single modulated carrier. This is called discrete multitone transmission (DMT).

DMT divides a high-speed data transmission signal into several lower speed parts; of which are carried on each sub-channel. When the sub-channels are received, the low speed parts are recombined to create the original high speed data transmission signal.

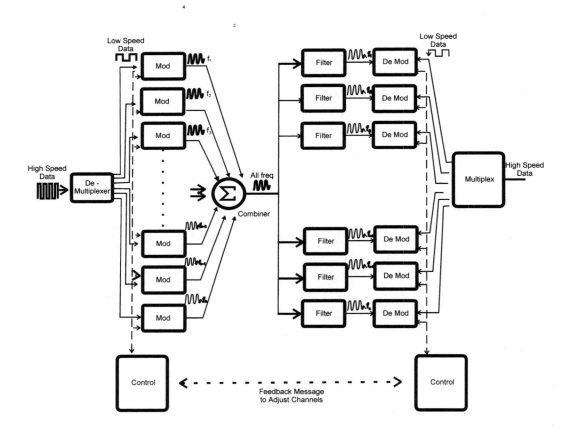

Figure 2.4, Discrete Multitone Transmission (DMT)

The advantage of sending several sub-channels is the ability to independently adjust the transmission levels of each sub-channel signal. Because the frequency response of the line can vary and distortions can occur on specific frequencies (where only a few sub-channels may be affected), DMT systems adapt well to the hostile environment of copper wire transmission.

Figure 2.4 shows discrete multitone transmission system (DMT). This diagram shows a high speed data signal is divided into several low speed signals. Each low speed signal modulates a sub-channel. The sub-channels are combined and supplied to the copper wire. At the receiving end, each sub channel is received and decoded. The sub-channel data signals are re-combined to recreate the original high-speed data signal.

Error Detection and Correction

As we have discussed previously, the transmission of information on the copper wire is subject to distortion from a variety of sources. This could be a brief distortion due to a high-voltage ring signal that leaks a signal to an adjacent pair of copper wires. The distortion could also be continuous. Such is the case when distortion occurs from crosstalk off a continuous digital transmission on an HDSL line that may be located in the same wire bundle.

When distortion occurs, it can cause some of the data to become corrupted or received in error. To help overcome the effects of distortion, error detection and protection bits may be added prior to transmitting the digital information. These extra bits of information allow the detection and possibly the correction of this information.

The disadvantage of sending extra bits is the loss of available data transmission capability and time delays for the data transmission. The more bits that are sent for error detection and correction, a lesser number of bits are available to carry data (such as a digital video signal). Unfortunately, the distortion caused on one copper line may be minimal, while the distortion on another line may be high. For example, a single line may be mounted on a pole in the country (low external noise) where another line may be located next to thousands of other low voltage and high voltage lines in a conduit pipe in the city (high external noise). Error detection and correction can also introduce time delays. This is due to the calculation periods for the error detection and correction formulas and distribution of error protection bits. These time delays typically vary from a few milliseconds (1/1000th of a second) to several seconds.

There are two basic types of error detection and protection used in DSL digital transmission; block coding and convolutional (continuous) coding. Block codes protect groups of information while convolutional codes protect information that is transmitted in sequence.

A block code is a series of bits, or a number that is added to a group of bits or batch of information, that allows for the detecting and/or correcting of information that has been transmitted. Block codes use mathematical formulas that per-

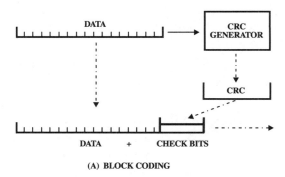

(A) BLOCK CODING

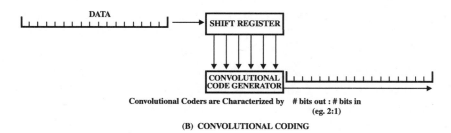

(B) CONVOLUTIONAL CODING

Figure 2.5, Error Detection and Protection Coding

form an operation on the data that will be transmitted. This produces a resulting number that is related to the transmitted data. Depending on how complex the mathematical formula is and how many bits the result may be, the bock code can be used to detect and correct one or more bits of information.

Convolutional coding is an error correction process that uses the input data to create a continuous flow of error-protected bits. As these bits are input to the convolutional coder, an increased number of bits are produced. Convolutional coding is often used in transmission systems that often experience short burst errors such as wireless systems. Convolutional coders are described by the relationship between the number of bits entering and leaving the coder. For example: a 1/2 rate convolutional coder generates two bits for every one that enters. The larger the input to output relationship, the larger the redundancy and this produces a more robust error protection. For example, a 1/4 rate convolutional coder has much more error protection capability than a 1/2 rate coder. However, the ¼ rate convolutional coder will produce 4 bits for every one bit of user data provided to the coder. If the communications channel capacity was 40 kbps, the user could only send 10 kbps.

Figure 2.5 shows how error detection and correction bits are added to a digital message. Error detection and correction bits add to the total number of bits used per subscriber. Therefore, error detection and correction reduces the number of bits available to users and decreases the system capacity.

Different types of error detection and protection processes are used in xDSL systems and their use sometimes depends on the type of service used by the customer. For example, errors that occur during digital voice communication or video broadcast may be acceptable; while errors that are received during a software program file transfer would be unacceptable. Thus, the customer may be willing to accept a higher error rate at the benefit of near real-time communications and higher bandwidth allowance.

Because there is a tradeoff of time delays and loss of bandwidth, most xDSL technologies offer at least two types of communications channels; fast and slow. Fast channels usually offer minimal error protection or slow data transmission rates. This allows for short delays in message transfer. Slow channels may use long interleaving periods or may have extra error protection. Slow channels can have delays in excess of a few seconds.

Protocol Adaptation

Protocol is the language and processes that are used to setup, maintain and disconnect a communications channel. Different types of systems use different types of protocols. When connecting systems of different types, protocols must be converted to allow for different command languages, timing issues and other key parameters.

For a DSL network that connects a computer to the Internet, there may be several protocol adaptations. This may include from USB to Ethernet, from Ethernet to ATM, and from ATM to TCP/IP. To complicate protocol conversions, some commands and timing sequences may not identically translate and the manufacturers implementation of protocol adaptation may be subject to interpretation. This can lead to incompatibility between different types of devices.

Figure 2.6 shows the necessity to convert the protocol from one system to another. This diagram shows that protocol adaptation contains conversion of messages from one protocol language to another. It also involves adapting timing and grouping of messages. This diagram shows that each time the protocols are adapted, it introduces timing delays. These timing delays can accumulate through a system, causing time out errors and other problems with end-to-end system communications.

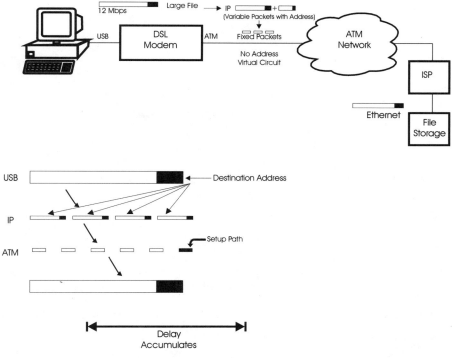

Figure 2.6, Protocol Adaptation

Transmission Medium Limitations

Some of the limitations of copper lines that reduce the ability to transfer digital information include limited frequency response of the copper lines, crosstalk, noise from external sources that cause distortion, non-terminated tap lines, audio loading coils, and signal attenuation that results from line splices and line resistance.

Frequency Response

The twisting of copper wire pairs provides good frequency response for low frequency audio signals. Unfortunately, twisted copper wire pairs are not specifi-

Figure 2.7, Frequency Response of Copper Wire

cally designed for high frequency transmission. Analog signals have a frequency range of up to 3.4 kHz and most of the xDSL technologies use frequencies up to 1.1 MHz. As the frequency applied to the copper wire pair increases, the attenuation of the line increases and signal energy leaks (emits) from the wire pair.

Figure 2.7 shows the typical frequency response of a twisted pair of copper wires. The frequency response depends on a variety of factors including the dimension of the copper wire (gauge), insulation type and installation environment (twisting or stapling of the wire).

Crosstalk (Signal Leakage)

Crosstalk is the undesired coupling of a signal from one communications channel to another. Crosstalk occurs when some of the transmission signal energy leaks from the cable. This leakage is called signal egress (emission from the line).

Crosstalk on DSL systems can be divided into two categories; near end crosstalk (NEXT) and far end crosstalk (FEXT). Figure 2.8 shows two types of DSL crosstalk. NEXT results when some of the energy that is transmitted in the desired direction seeps into one (or more) adjacent communication lines from the originating source. FEXT occurs when some of the digital signal energy leaks from one twisted pair and is coupled back to a communications line that is transferring a signal in the opposite direction. Generally, NEXT is more serious than FEXT as the signal interference levels from NEXT are higher.

Figure 2.8, FEXT and NEXT Crosstalk

Signal Ingress

Signal ingress occurs when electrical signals from other sources (such as radio or lightning spikes) enter into the transmission line. Figure 2.9 shows several sources of radio signal ingress that may occur in a DSL system. This diagram shows that a high power AM radio transmission tower that is located near a telephone line couples some of its energy onto the telephone line. This interference signal (radio ingress) usually reduces the data transmission capacity of the DSL line.

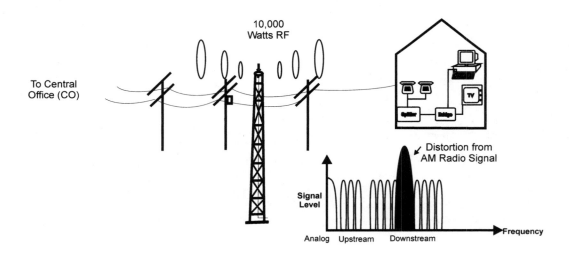

Figure 2.9, Radio Signal Ingress

Bridge Tap Reflections

A bridge tap is an extension to a communication line that is used to attach two (or more) end points (user access lines) to a central office. Bridge taps provide connection options to the telephone company on connecting different communication lines to a central office without having to install new pairs of wires each time a customer requests a new telephone line.

The connection of one (or more) bridge taps on a communication line that is used for POTS service does not usually cause signal distortion. However, unterminated bridge taps that are installed on communication lines that transfer xDSL signals can result in signal distortion. The signal distortion comes from the reflections of signal energy reflections off the bridge taps.

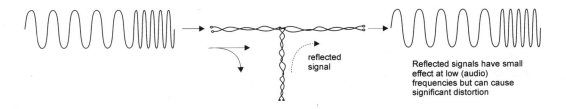

reflected
signal

Reflected signals have small
effect at low (audio)
frequencies but can cause
significant distortion

Figure 2.10, Bridge Tap Reflections

When an electrical signal is applied to the end of a copper wire, electrical energy begins to travel down the copper wire. Ideally, when the energy reaches the end of the copper wire, the signal is absorbed at the other end (called a matched line). If the end of the wire is not connected, some (or all) of the energy is reflected back to the beginning of the line.

Figure 2.10 shows how reflections from bridge line tap can cause distortion. This signal shows that some of the energy from the bridge tap is reflected back to the communications line. This reflected signal is a delayed representation of the original signal. Typically, bridge taps must be removed from communications lines that use DSL technology.

Audio Loading Coils

Audio loading coils are used to adjust the frequency response of a communication line to better transfer audio signals. While these loading coils work well for audio signals, they completely disable the ability of the line to be used for high frequency xDSL signals.

Figure 2.11 shows that there may be several installed audio loading coils on a single local loop line. These loading coils must be removed for xDSL transmission.

Line Splice Attenuation

Telephone cables usually come in 500 feet roles. Because most telephone lines are several thousand feet from the central office, several cable splices are required. Each line splice attenuates the signal and the amount of signal attenuation varies depending on the type of splice (solder, twist or pegs) and the amount of corrosion inside the splice.

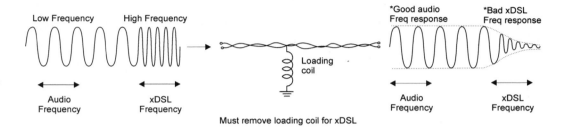

Figure 2.11, Audio Loading Coils

Because the average distance for local access lines is over 10,000 feet, there are more than 20 splices in the average local access loop. Each of these splices offers the potential for corrosion and increased resistance.

One method that is used to decrease the effects of corrosion (and reduce the attenuation) is to continuously run electric current through the copper wire pair. This "sealing current" is a small amount of direct current that is passed through a copper wire to reduce the corrosion effects of the splice points. The sealing current effectively maintains conductivity of mechanical splices that are not soldered. The direct current effectively punches holes in the corrosive oxide film that forms on the mechanical splices.

Line Resistance Attenuation

The copper cable also has resistance (impedance) that is dependent on the size (diameter) of the cable. The resistance of the copper wire increases as the diameter decreases (gauge number increases). The higher the line resistance, the more of the signal energy is dissipated by the line and less energy is transferred to the receiving device. Figure 2.12 shows how line resistance attenuation and the wire size decreases.

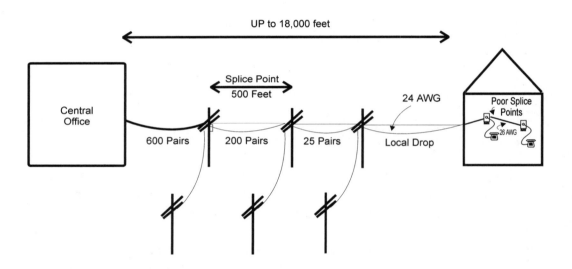

Figure 2.12, Line Resistance Attenuation

References:

1. Albert Azzam, "Broadband Access Technologies," McGraw-Hill, New York, 1998, pg. 106.

Chapter 3

Introduction to xDSL

xDSL systems evolved from the necessity to transfer digital information for the last few thousand feet of the local loop. There are several different xDSL technologies. The earlier xDSL technologies (e.g. HDSL) were digital only systems that replaced older digital transmission technology. The newer xDSL systems allow for simultaneous analog and digital signal transmission at much higher data rates.

The technology behind xDSL systems continues to evolve. Although there are standards, various proprietary systems have been developed for xDSL systems that deviate from the standards. The result is some xDSL equipment is not compatible with other xDSL equipment of the same classification.

History of xDSL

The first digital subscriber lines (DSLs) were developed due to the need for cost effective quality communication over copper wire. The first digital transmission system was the T1 line. This system had a maximum distance of approximately 6,000 feet prior to needing repeaters.

Figure 3.1 shows the evolution of DSL systems. The first xDSL system was high speed digital subscriber line (HDSL). The HDSL system increased the distance that high speed digital signals could be transmitted without the user of a

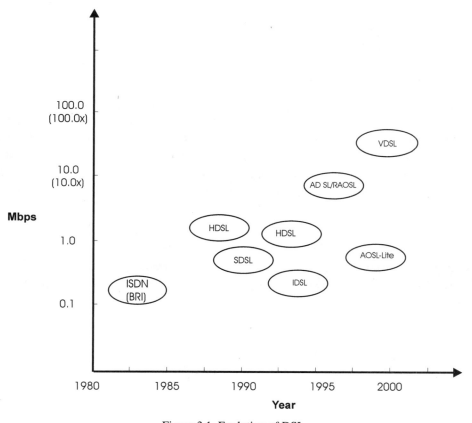

Figure 3.1, Evolution of DSL

repeater/amplifier. The HDSL system did require 2 (or 3) pairs of wires to allow simultaneous (send and receive) up to 2 Mbps of data transmission. To conserve the number of copper pairs for data transmission, synchronous digital subscriber line technology was developed. Although SDSL systems offered lower data rates than HDSL, only 2 wire pairs were required. Since SDSL was developed, the HDSL system has evolved to a 2nd generation (HDSL2) that allows the use of 2 wire pair for duplex transmission with reduced emissions (lower egress). New efficient modulation technology used by ADSL systems dramatically increased the data transmission rates from the central office to the customer to over 6 Mbps (some ADSL systems to 8 Mbps). To take advantage of ISDN equipment and efficiency, an offshoot of ISDN technology that was adapted for the local loop developed called IDSL. ADSL systems evolved to rate adaptive digital subscriber line (RADSL) allow the data rate to be automatically or man-

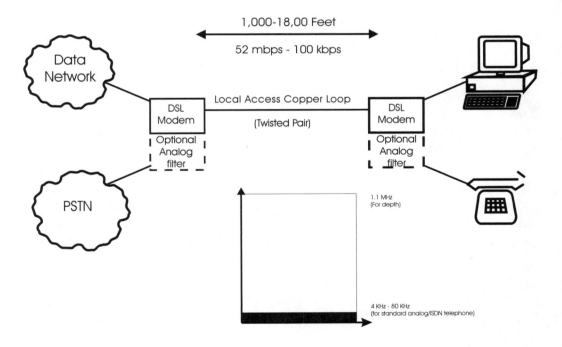

Figure 3.2, Basic xDSL System

ually changed by the service provider. To simplify the installation of consumer based DSL equipment, and low data transmission offshoot of ADSL developed that is called ADSL-Lite. Using similar technology as the ADSL system, very high speed digital subscriber line (VDSL) was created to provide up to 52 Mbps data transfer rates over very short distances.

Figure 3.2 shows a basic xDSL system. This diagram shows that the key to xDSL technologies is a more efficient use of the 1 MHz of bandwidth on a single pair of copper telephone lines. An xDSL system consists of compatible modems on each end of the local loop. For some systems, the xDSL system allows for multiple types of transmission on a single copper pair. This includes analog or ISDN telephone (e.g. POTS) and digital communications (ADSL or VDSL). This diagram shows that there are basic trade offs for DSL systems. Generally, the longer the distance of the copper line, the lower the data rate. Distances of less than 1,000 feet can achieve data rates of over 50 Mbps.

xDSL Technologies

There are many types of digital subscriber line (DSL) technologies and these technologies continue to evolve. These include IDSL, HDSL, SDSL, HDSL2, ADSL, RADSL, CDSL and VDSL. ISDN digital subscriber line (IDSL) is a simplified version of basic rate ISDN lines. High speed digital subscriber line (HDSL) was the first digital subscriber line technology that requires at least 2 copper wire pairs. Symmetric digital subscriber line (SDSL) was a 2 wire version of HDSL. High speed digital subscriber line 2 is an advanced 2-wire version of standard HDSL2 (different than SDSL). Asymmetrical digital subscriber line (ADSL) was the first DSL technology that had different transfer speeds for the forward and reverse direction. Rate adaptive digital subscriber line (RADSL) is a version of ADSL that allows for the automatic changing of data transmission rates. Consumer digital subscriber line (CDSL) is a version of ADSL that simplifies the network hardware through a reduced data transmission rate. Very high speed digital subscriber line (VDSL) is a very high speed (up to 52 Mbps) DSL technology that is used for very short distances.

ISDN Digital Subscriber Line (IDSL)

ISDN Integrated services digital network (IDSN) is a digital technology that precedes xDSL systems. Similar to xDSL technology, ISDN also works over standard copper phone wiring. Each ISDN line provides two digital channels at 64 kilobits per second and one digital channel (used primarily for control purposes) at 16 kbps. Although ISDN has gained some popularity in Europe, its deployment is less than 5% worldwide.

ISDN Digital Subscriber Line (IDSL) is a hybrid of ISDN and DSL technologies. It uses the same data formatting as ISDN devices on the copper wire pair and delivers up to 144 kilobits per second bandwidth through two 64 kbps channels and one 16 kbps channel. The key difference for IDSL systems is that the IDSL system only uses the 64 kbps DS0 channels and the ISDN control channel (D channel) is ignored. The IDSL system effectively multiplies the number of channels on a single copper pair by 2x. The ability to avoid using ISDN signaling is very important as software upgrades for switching systems, to allow ISDN operation can cost more than $500,000 per switch.

High Bit Rate Digital Subscriber Line (HDSL)

HDSL technology was developed to overcome some significant limitations of the first digital transmission systems (T1 and E1). These limitations included a maximum distance of 6,000 feet between repeaters and a requirement for line conditioning.

Early electronic processing systems for digital transmission technology used in the 1960's and early 1970's were expensive. However, in the 1980's, advanced low cost signal processing systems became available. This allowed more efficient data transmission technologies to be developed and cost effectively deployed.

For T1 and E1 systems, the physical characteristics of copper lines were modified to allow digital transmission (line leveling). Because it required equipment and trained personnel, this was an expensive process. It was discovered that the high speed digital technology that was developed in the 1960s for T1 (1.544 Mbps) and E1 (2 Mbps) transmission could be replaced with more efficient transmission technologies that did not require quality line conditioning. By allowing the electronics to adjust for the line conditions, this permitted for more rapid and cost effective installation of high speed data transmission circuits.

Figure 3.3 shows a basic HDSL system. This diagram shows that the first application for HDSL used two pairs (and sometimes 3 pairs) of copper wire. Each

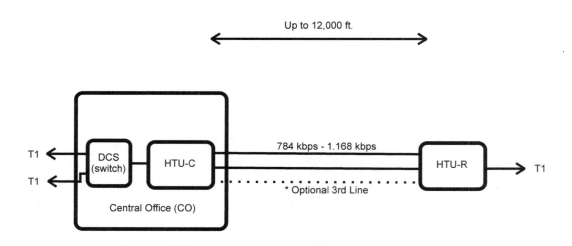

Figure 3.3, High data rate Digital Subscriber Line (HDSL)

circuit has an HDSL Termination Unit (HTU) on each end; an HTU-C (central office) and HTU-R (remote). This example shows that each pair of HDSL wires carries 784 kbps full duplex (simultaneous send and receive) data transmission. To carry the equivalent of a T1 line, two pairs of lines are used. It is also possible to carry an E1 line by using 3 pairs of copper wire. Although the framed transport for HDSL is different than for a T1 or E1 line, the HTU-C and HTU-R convert the protocols to standard T1 lines.

The key advantage to the HDSL system is the increase in distance a line may go between repeaters. HDSL systems can go up to 12,000 feet (or more with heavier wire gauge) where as a T1 or E1 line can only go to 6,000 feet between repeaters. HDSL lines are also more tolerant to bridge line taps than T1 or E1 lines. At the end of 1999, most of the new T1 leased lines in the United States were provided by using HDSL technology.

Symmetric Digital Subscriber Line (SDSL)

Symmetric Digital Subscriber Line (SDSL) offers an equal (symmetric) transmission rate of data at the same speed as HDSL, with two key differences. SDSL can be offered using only one copper pair (as opposed to 2 pair required for HDSL) and the line distance must be less than 10,000 feet from the phone company's central office.

High Bit Rate Digital Subscriber Line - 2nd Generation (HDSL2)

HDSL second generation (HDSL2) was developed in 1998 as the next generation of HDSL technology. The key features of the HDSL2 system include minimized interference to other lines in cabling bundles, extended range, increased data transmission rate and the ability to provide data transmission on a single pair of wires. To achieve these objectives, the line coding (modulation type) used for HDSL2 system differs from its predecessors (HDSL and SDSL).

Asymmetric Digital Subscriber Line (ADSL)

Asymmetric Digital Subsdriber Line (ADSL) is a communication system that transfers both analog and digital information on a copper wire pair. The analog information can be a standard POTS or ISDN signal and the maximum down-stream digital transmission rate (data rate to the end user) can vary from 1.5 Mbps to 9 Mbps downstream and the maximum upstream digital transmission rate (from the customer to the network) varies from 16 kbps to approximately 800 kbps. The data transmission rate varies depending on distance, line distor-tion and settings from the ADSL service provider. ADSL technology is capable of transmitting digital movies, television, picture catalogs, CD quality audio, links for high-speed corporate networks, and high-speed Internet small busi-nesses and homes.

Unlike early xDSL technologies, ADSL is asymmetric in nature. It was discov-ered that the many customers only desired services that required high-speed data transmission from the network. These services do only required to send low-speed or moderate-speed data back through the network. The ADSL transmis-sion channel can be divided into several high speed data channels and at the same time continue to provide standard POTS telephone service. The standard POTS or ISDN channel is separated from the digital modem by filters. This allows uninterrupted POTS or ISDN even if the high-speed data transmission system becomes inoperative.

The downstream channels can be set from 1.5 to 6.1 Mbps (up to 8 Mbps on some systems) and the upstream data transmission rates range from 16 to 640 kbps (up to 800 kbps on some systems). Each data channel can be further sub divided (sub-multiplexed) to form multiple, lower speed channels as necessary (e.g. one channel for digital television and one channel for Internet web surfing).

The maximum data transmission rate for ADSL modems varies dependent on the distance, interference levels, presence of bridge taps, quality of the copper line and setting provided by the ADSL service provider. Generally, ADSL will pro-vide the following data transmission rates shown in Figure 3.4.

Data Rate	Wire Gauge	Distance
1.5 or 2.0 Mbps	24 AWG/0.5 mm	18,000 feet/5.5 km
1.5 or 2.0 Mbps	26 AWG/0.4 mm	15,000 feet/4.6 km
6.1 Mbps	24 AWG/0.5 mm	12,000 feet/3.7 km
6.1 Mbps	26 AWG/0.4 mm	9,000 feet/2.7 km

Figure 3.4, ADSL Data Transmission Rates

Source: DSL Forum

Various protocols are available for ADSL modems including Ethernet, ATM and Internet Protocol. This allows the customer an almost plug and play capability, if DSL service is available in their area.

The G.dmt version of ADSL can transfer data downstream at rates up to 8 Mbps and transfer data upstream at rates up to 1.5 Mbps. If the modem is located at a distance more than 10,000-12,000 feet from the central office (CO), the data transmission rates decrease. G.dmt can provide up to 1.5 Mbps of data transmission up to 18,000 feet from the central office. If standard telephone devices are used on the same line, a "splitter" must be installed on the phone line to separate the analog signal from the high-speed digital signal.

According to many industry experts, over 95% of local access loop copper pairs have a distance that is within the reach of ADSL service. Customers that are located beyond these distances can be reached with fiber-based digital loop carrier systems. Unfortunately, some local telephone companies use the integrated digital loop carrier system (IDSL) for digital transmission in the local access network. The IDLC system is not directly compatible with ADSL service.

Initially, an ADSL Forum was established to help promote ADSL services and equipment. Because the ADSL Forum has been involved with the evolution of DSL technology (e.g. VDSL and ADSL-Lite), it has been renamed DSL Forum.

The first official line coding (modulation and data structure) for ADSL systems was DMT. Since its introduction, variants that use CAP line coding have emerged. It is likely that variations of line coding will continue to emerge improving the performance of ADSL systems.

Figure 3.5 shows a typical ADSL system diagram. This diagram shows that a single copper access line can be connected to different networks. These include the public switched telephone network (PSTN) and the data communications network (usually the Internet or media server). The ability of ADSL systems to

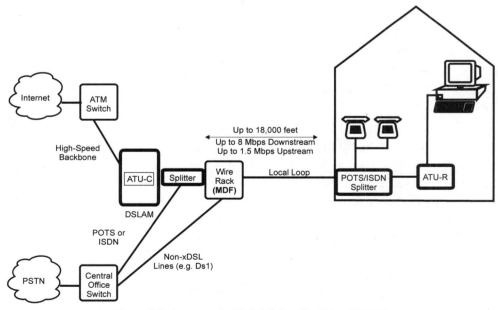

Figure 3.5, Asymmetric Digital Subscriber Line (ADSL)

combine and separate low frequency signal (POTS or IDSN) is made possible through the use of a splitter. The splitter is composed of two frequency filters; one for low pass and one for high pass. The DSL modems are ADSL transceiver unit at the central office (ATU-C) and the ADSL transceiver unit at the remote home or business (ATU-R). The digital subscriber line access module (DSLAM) is connected to the access line via the main distribution frame (MDF). The MDF is the termination point of copper access lines that connect end users to the central office.

Rate Adaptive Digital Subscriber Line (RADSL)

Rate Adaptive Digital Subscriber Line (RADSL) operates at the same bandwidths as ADSL with the added capability of dynamically changing bandwidth. The bandwidth may change due to the quality of the phone line during the data transmission or as a result on limitations imposed by the service provider (different rates for different services).

Consumer Digital Subscriber Line (CDSL)

Consumer Digital Subscriber Line (CDSL) or Universal ADSL was developed to overcome a key installation challenge of ADSL systems. CDSL is also known as ADSL-lite or G.Lite. CDSL eliminates the requirement of having to install a splitter in the home or business. However, the key tradeoff is a lower maximum data transmission rate (approximately 1.5 Mbps downstream and 384 kbps upstream maximum).

Figure 3.6 shows a CDSL system. This diagram shows a typical CDSL system diagram. This diagram shows that the basic DSL network is similar to the ADSL network. The primary difference is in the end user equipment and its connection to the telephone network. The CDSL system does not require a splitter for the home or business. Instead, the end user can install microfilters between the tele-

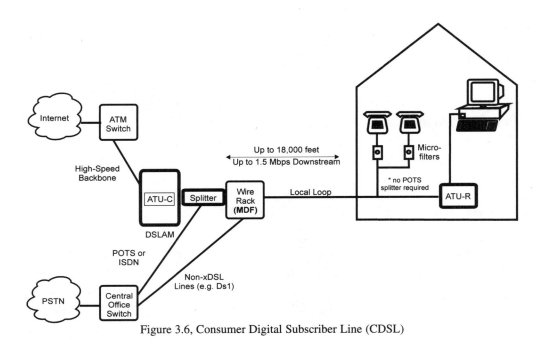

Figure 3.6, Consumer Digital Subscriber Line (CDSL)

phone line and standard telephones. These microfilters block the high speed data signal from interfering with standard telephone equipment. The CDSL end user modem contains a filter to block out the analog signals.

Very High Bit Rate Digital Subscriber Line (VDSL)

At the beginning of 2000, Very High Bit-Rate Digital Subscriber Line (VDSL) was the fastest DSL technology available. VDSL offers rates from 13 to 52 megabits per second downstream and 1.5 to 26 megabits per second upstream. Unfortunately, such high data transmission rates are possible at limited distances (approximately 1,000 and 4,500 feet).

VDSL downstream transmission rates were derived from data transmission rates associated with ATM, SONET and SDH. These data rates include 51.84 Mbps, 25.92 Mbps and 12.96 Mbps. Figure 3.7 shows the approximate distance and data rates associated with VDSL.

VDSL is seen as an enabling technology for getting fiber optic data speeds closer to the home or business. VDSL allows use of analog telephones along with high speed data connections. However, VDSL only transfers high speed data for short distances.

VDSL technology resembles ADSL technology. However, because ADSL was designed to adapt to more hostile line conditions than VDSL systems are expected to, VDSL technology is actually much simpler to implement.

The initial versions of VDSL use frequency division multiplexing to separate the upstream and downstream channels. However, new versions of VDSL can share

Downstream Data Rates	Maximum Distance
12.96 - 13.8 Mbps	4500 feet/1.5 km
25.92 - 27.6 Mbps	3,000 feet/1.0 km
52.84 - 55.2 Mbps	1,000 feet/0.3 km

Upstream Data Rates
1.6 - 2.3 Mbps
19.2 Mbps
Equal to Downstream

Figure 3.7, VDSL Data Transmission Rates

Source: DSL Forum

the same frequency bands. When the same frequency bands are used, echo canceling is required.

VDSL uses the frequency above ADSL where amateur radio operates. Telephone wires act as antennas that can capture and transmit these radio signals. In short, to use an amateur radio operation near VDSL lines can dramatically decrease the data transmission capability of VDSL systems. VDSL transmission can also interfere with radio signals.

Figure 3.8 shows a VDSL system. This diagram shows that a fiber network reaches a neighborhood or small group of buildings. The fiber terminates in an optical network unit (ONU). The ONU converts the optical signal into an electrical signal that can be used by the VDSL modem in the DSLAM. The DSL modem signal is supplied to a splitter that combines the analog and digital sig-

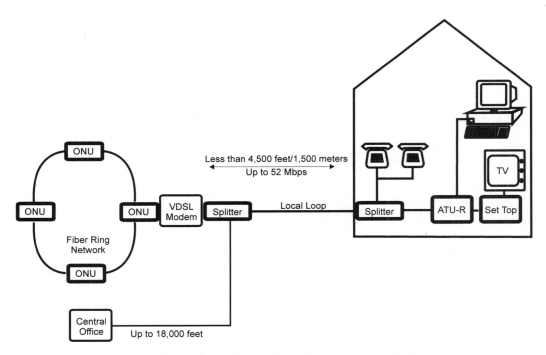

Figure 3.8, Very high bit rate Digital Subscriber Line (VDSL)

nal to copper access line. The splitter is actually attached to the last few hundred feet of the copper access line. The figure shows that the analog POTS signal from the local telephone company may still travel thousands of feet back to the central office. At the customers premises, the VDSL signal arrives to a splitter that separates the analog signal from the high speed digital VDSL signal. Because VDSL has a much higher data transfer rate, the CPE may include a digital video set top box that allows for digital television.

Technology Comparison Summary

Generally, the availability of new modulation technology and low cost electronic circuits that can do advanced signal processing (e.g. echo canceling and multiple channel demodulation) has enabled a dramatic change in the ability of copper wire to deliver high speed data signals.

DSL Technology	Down Stram Data (kbps)	Upstream Data (kbps)	Pots/ISDN Co-existence	Max Distance (feet)
IDSL	128	128	No	18,000
HDSL*	784-2048	784-2048	No	12,000
SDSL	384	384	No	12,000
HDSL 2	1,544-2048	1,544-2048	No	12,000
ADSL/ RADSL	1,000-8,000	1,000-8,000	Yes	18,000
ADSL Lite	800-2,000	16-200	Yes	18,000
VDSL	1,000-52,000	1,000-52,000	Yes	4,500 (1,000 for highest rates)

* Note Requires 2 to 3 wire pairs

Figure 3.9, xDSL Technology Comparison Summary

Source: DSL forum

Each of the xDSL technologies were developed to satisfy a specific business need and technology challenges. The key business drivers for xDSL technology development includes ability to have backward compatibility (simultaneous analog and digital capability), high speed data transmission rate, ability to operate on local loop lines with poor cable conditions, maximum distance without repeaters and simple installation.

Figure 3.9 shows a comparison of the different xDSL technologies. This table shows that tradeoffs exist between each of the technologies. Generally, the longer the distance, the lower the data rate. However, with advanced signal processing, higher data rates can be achieved at longer distances.

Chapter 4

Other Broadband Technologies

There are other competing broadband technologies that offer high speed broadband data transmission rates to the home or business. These other broadband communications systems include coax cable television (CATV) systems, optical systems (Fiber), fixed and mobile wireless systems and electric powerline technology.

Coax cable television systems are converting from one-way shared data systems to high bandwidth hybrid fiber coax systems. Optical systems have been successfully tested since the late 1980's. However, the high cost of deploying optical fiber to the home has limited its application in the residential marketplace. Wireless systems can be divided into narrowband (cellular and PCS) and wideband (wireless cable). Although wireless systems are technically capable of providing broadband services, the high value of mobility compared to bandwidth limits its use primarily to low bandwidth voice services. As of 2000, several electric powerline data distribution systems were in an early deployment stage.

The market for telephone service throughout the world is growing faster than 30% per year. Of the 6+ billion people in the world in 2000, there were only 700 million telephone lines. The number of telephone lines is projected to grow to 1.2 billion by 2002 [1]. In 1998, the marketplace for fixed telecommunications services in the United States was estimated at $110 billion, with local telephone and data services accounting for approximately $47 billion [2]. Fixed wireless may be used to provide customers with a broad range of telecommunications services that to bypass the established local exchange carriers (LECs). Predictions show that by the year 2005, wireless local loop is expected to soar to around 33 million lines in the United States.

Cable Television System

Cable television systems were initially designed to broadcast analog video signals in one direction to the end customer. Since the late 1980's, cable television systems have been undergoing a change from one-way analog broadcast to two-way digital transmission. Cable systems that offer cable modem technology are one of the primary competitors to DSL systems.

In the mid-1990's, cable television service providers ("cable companies") began to offer cable modems that allowed customers to access the Internet. Cable systems offer hundreds of high-bandwidth television channels where each television channel has 6 MHz of available bandwidth. Compare this to a single pair of copper wires that is used for DSL transmission. Although it is possible to use copper wire pairs to transfer over 6 MHz of bandwidth, this is only possible over very short distances, typically less than 1,000 feet. In practical applications, each copper wire pair only has a usable bandwidth of slightly over 1 MHz for typical local access line distances of 5,000 feet or more.

A single television channel that is converted to a digital channel can provide up to 30 Mbps of data transmission. A single coaxial cable has approximately 600-800 MHz of bandwidth. Of this, the upper portion of the frequency range (550-

750 MHz) is often used for data transmission. This upper frequency band can provide over 1 Gbps of data.

The coaxial cable used in cable television distribution systems is a high quality medium compared to the often changing transmission medium of twisted pair copper lines. As of the end of 1999, over 2 million cable modems had been installed in the United States [3]. Cable television networks provide reasonably reliable service and usage cost for cable modems range from $40-60/month [4]. Of the cable modems that have been deployed, most are used for residential Internet access, as office buildings do not usually have cable television lines installed.

There are two basic types of cable modems; first generation one-way cable modems and the second generation of two-way hybrid fiber/coax (HFC) systems. One-way modems offer up to 2 Mbps download speed while HFC systems currently offer up to 30 Mbps.

First Generation (One-Way) Cable Modems

First Generation one-way cable modems transmit high-speed data to all the users into a portion of a cable network and return low speed data through telephone lines. Until the 1980's, most community access television (CATV) system used primarily one-way analog video broadcast technology. In the1990's, cable television systems began to offer telephone and Internet service. Although one-way cable modems can offer peak download transmission speed up to 30 Mbps, the shared use of these data signals limits the user's data transmission speeds range from 400 kbps to 1.4 Mbps [5]. The one-way cable modem approach is viewed as a progressive step used by cable companies while they upgrade to high-speed data two-way infrastructure.

Figure 4.1 shows a community access television (CATV) system that offers a one-way modem. In this diagram, data replaces a television channel and is sent

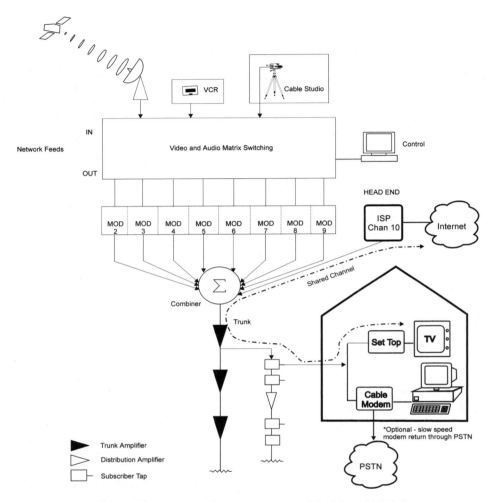

Figure 4.1, One-Way Modem in a Community Access Television (CATV) System

to all users in the network (or portion of the network that has the service). Each user in the network only can decode the data that is designated for them. When the user wants to transmit data, it is sent via the public telephone network at slower data transmission speeds.

Hybrid Fiber Coax (HFC)

The hybrid fiber coax (HFC) system is an advanced CATV transmission system that uses fiber optic cable for the head end and feeder distribution system and a coax for the customers end connection. HFC are the 2nd generation of CATV systems. They offer high-speed backbone data interconnection lines (the fiber portion) to interconnect end user video and data equipment. Many cable system operators anticipating deregulation and in preparation for competition began to upgrade their systems to Hybrid Fiber Coax (HFC) systems in the early 1990's. As of late 1999, approximately 35% of the total cable lines in the United States had already been converted to HFC technology [6].

The key differences between standard CATV and HFC networks include additional head end equipment for data communications, the conversion of trunks and distribution cables to fiber and the addition of two-way access nodes at the end of the network that route data signals to and from end customers. The cost of converting coax to fiber into the trunk and distribution part of the CATV network is small as fiber cable can simply be attached to existing coax cable on telephone poles.

Figure 4.2 shows a hybrid fiber coax (HFC) system. In this diagram, the head end of the system is connected to media sources in a similar manner to the standard tree structure that is used in earlier CATV systems. The HFC head end usually includes a data network server (typically an ISP server and a gateway to Internet). The HFC feeder distribution system uses fiber optic cable that can carry many video and high-speed data channels. These optical signals are distributed through hubs to provide service to neighborhoods and business centers. The hubs eventually terminate into fiber nodes. The fiber nodes convert the video and data signals into modulated channels that are transmitted on the last few kilometers of coax cable.

The HFC fiber distribution system offers the potential for two-way communications. This allows end users to obtain high-speed connections to the Internet without the need to share a common channel with other users. As shown in

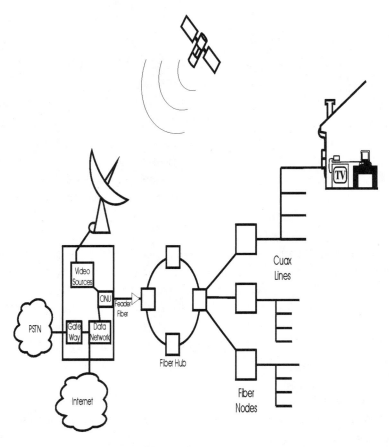

Figure 4.2, Hybrid Fiber Coax (HFC) System

Figure 4.2, it is also possible for HFC systems to provide for voice communications. This would require the CATV provider to have a gateway to the PSTN and for the fiber nodes to accept phone service from the end customer's equipment. This would require digital telephones (e.g. personal computers) or a special adapter that allows digital channels to convert to analog telephones.

Wireless Cable

"Wireless Cable" is a term given to land based (terrestrial) wireless distribution systems that utilize microwave frequencies to deliver video, data and/or voice signals to end-users. Wireless cable provides video programming from a central location directly to homes via a small antenna that is mounted on the side of the house. There are two basic types of wireless cable systems, Multichannel Multipoint Distribution Service (MMDS) and Local Multichannel Distribution Service (LMDS). In 1998, there were over 10 million MMDS wireless cable customers throughout the world and over 1.1 million in the United States.

In the 1970's, the first Multipoint Distribution Service (MDS) was the direct delivery of Home Box Office (HBO). Later, HBO moved to satellite transmission. The FCC reallocated many of these MDS frequency channels in the early 1980s and officially renamed them MMDS.

In 1996, some analog MMDS systems began upgrading to digital service. Through the use of digital video compression, digital transmission allows five to six times the video channel capacity. In addition to video programming, wireless cable can provide telephone service and data services.

Figure 4.3 shows that the major component of a wireless cable system is the head-end equipment. The head-end equipment is equivalent to a telephone central office. The head-end building has a satellite connection for cable channels and video players for video on demand. The head end is linked to base stations (BS) which transmits radio frequency signals for reception. An antenna and receiver in the home converts the microwave radio signals into the standard television channels for use in the home. As in traditional cable systems, a set-top box decodes the signal for input to the television. Low frequency MMDS wireless cable systems can reach up to approximately 70 miles. LMDS systems can only reach approximately 5 miles.

Wireless cable is one of the most economical technologies available for the delivery of pay-per-view television service. Wireless cable systems do not

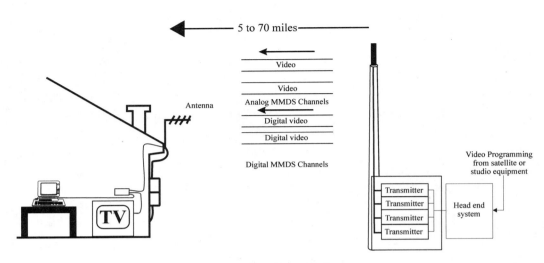

Figure 4.3, Wireless Cable System

require extensive networks of cables and amplifiers, bringing the offered price generally lower than a traditional cable service. To the customer, a wireless cable system operates in the same manner as a traditional cable system. Because wireless signals are transmitted over the air rather than through underground or above-ground cable networks, wireless systems may be less susceptible to outages, offer better signal quality and are less expensive to operate and maintain than traditional cable systems. In conventional coaxial cable distribution networks, the television signal quality declines in strength as it travels along the cables, and must be boosted by amplifiers, thus introducing distortion into the television signal.

To add security for wireless cable systems, so unauthorized users do not gain access to the system (stealing service), signals from video sources are scrambled with a code. The user must have the code to successfully view the video signals. Like traditional cable systems, wireless cable systems employ "addressable" subscriber authorization technology, which enables the system operator to control centrally the programming available to each individual subscriber, such as a pay-per-view selection.

There are two primary methods of providing a communication path back from the end customer to the network operator; a telephone line and wireless. Wireless cable systems have commonly only provided wireless downlink service (radio transmission from the system to the customer). Some of the new wireless cable systems now dedicate some of their radio channel capacity to uplink channels (from the customer to the system). This allows wireless cable systems to offer two-way service. Uplink channels typically are used to allow the customer to select programming sources (such as pay per view) or provide for two-way Internet access. Two-way service can also provide telephone service.

Multmegabit Multimedia Distribution System (MMDS)

MMDS is a wireless cable service that is used to provide a series of channel groups, consisting of channels specifically allocated for wireless cable (the "commercial" channels). In the United States, MMDS service evolved from radio channels that were originally authorized for educational purposes. MMDS video broadcast systems have been in service since the early 1990's providing up to 33 channels of analog television over a frequency range from 2.1 to 2.69 GHz. Optionally, there are 31 "response channels" available near the upper end of the 2.5 to 2.69 band. These response channels were originally intended to transmit a voice channel from a classroom to a remote instructor.

MMDS systems normally transmit using an omni-directional antenna with the receivers, usually at a home, using a small directional antenna. A MMDS service provider typically supplies a frequency down-converter (set top box) that converts the MMDS frequencies to ordinary VHF TV channels.

Although MMDS wireless cable is a relatively established service, FCC rule changes in 1990 allowed single operators to license larger numbers of channels and operate them as a broadcast-like service, which made MMDS more economically viable. After this change in rules, a tidal wave of industry interest resulted in 24,000 license applications. Well over 160 MMDS systems were operating in the United States in 1997, with half of them serving rural areas. It is possible that "private" educational video transmissions will be widely distributed by optical fiber networks, rather than by MMDS services, keeping MMDS service offerings in check. MMDS is currently used to enrich junior college classroom offerings and used for adult evening classes at remote locations.

Local Multipoint Distribution Service (LMDS)

Local multipoint distribution service (LMDS) provides similar service as MMDS at higher frequencies. Because of the use of high frequency that is attenuated quickly, LMDS must install antennas approximately every 5 miles. When antennas are installed, this offers tremendous system capacity possibilities. Digital LMDS has the capacity for thousands of video, audio and data channels. MMDS and LMDS both will enable wireless cable systems to transmit high definition television (HDTV) over their existing allocated frequencies and channels. LMDS has more than twice the bandwidth of AM/FM radio, VHF/UHF television, and cellular telephone combined. LMDS may prove to be the shortest path to mass delivery of two-way video and high-speed data services ever commercially offered.

Several LMDS systems are in operation. These systems are analog and several digital systems are in development. CellularVision, a company that provides service to New York City has been providing service since 1996. Recently, Cellular Vision offered wireless Internet service to consumers that allowed data rates of 500 kbps. CellularVision also announced that it plans to provide wireless Internet to businesses with data rates up to 32 Mbps. Other companies that are developing LMDS technology include Hewlett Packard, Texas Instruments, Stanford Telecom and others.

Initially, radio coverage was perceived to be a challenging factor for companies that offer LMDS service because microwave signals cannot regularly penetrate into buildings very well. This would have limited the deployment of LMDS systems into large cities such as New York City. However, because microwave signals reflect off large objects (such as buildings), wireless cable companies have been exploiting the reflections off buildings to provide extended coverage.

The key challenge for LMDS operators will likely be the initial construction of systems. Unlike cellular telephone operators, which were able to start with large cells and gradually split them into smaller cells, LMDS systems can only offer small radio coverage areas (cells). This means that LMDS service providers will likely target areas with a large concentration of potential users.

LMDS frequencies in the United States are near 28 and 29 GHz downlink and near 31 GHz for an uplink return path (transmission from the end customers equipment to the system operator).

Wireless Local Loop (WLL)

Wireless local loop (WLL) generally refers to the use of wireless technology to provide voice, data, or video service to fixed locations. There are several fixed wireless systems that can replace or bypass services that have traditionally been provided by copper wire or fiber cable. Wired systems that may be replaced or bypassed include wired telephone service, high-speed telephone communication links, cable television systems and local area network systems.

When the Telecommunications Act of 1996 was passed by the United States Congress, the local telephone monopoly in the United States was expected to end, just as it did in the United Kingdom in early 1990s. Like the UK and the US, many countries are now deregulating telecommunications (PTT monopolies) by allowing new competition for local telephone, long distance telephone and cable services. For example, India and Indonesia will soon have wireless local loop providers competing with the existing telephone company.

The terminology of "local telephone service" and "cable service" as we know it is changing. Already some cable companies are offering local telephone service and some telephone companies are offering television services. New companies are being formed to compete with the local telephone and local cable companies. Many of these new entrants will use wireless as their access to the local loop. The use of wireless allows rapid deployment of services and reduces the cost of installing cables to each residence or building.

Using wireless systems instead of wired systems allows new entrants to keep the system construction costs down while deploying the systems quickly (in months instead of years in some cases). The basic fixed wireless technologies that are being introduced to replace or bypass cables include wireless local loop (WLL), wireless cable, wireless bypass, and wireless local area networks (WLAN). These fixed wireless services can provide local dialtone voice service, high-speed data and video service. In some cases, a single fixed wireless system may provide all these services at the same time.

The demand for wireless local loop is segmented into two key areas: services for developed countries and services for undeveloped countries. Developing markets are showing interest due to the demand for basic telephone service. Developing markets are expected to significantly outpace developed markets for WLL service. It is estimated that by the year 2005, developing markets will have 148 million WLL subscribers while developed markets have only 54 million WLL subscribers.

In developed countries, it takes value added services and a tie to mobility services to get interest in WLL. In the particular case of the United Kingdom, point-to-multipoint (PMP) wireless service made some inroads by providing high quality WLL for Internet.

The demand for WLL in developing countries is projected to be concentrated in China, India, Indonesia, Brazil, and Russia. These five markets are projected to account for nearly three-quarters of demand by developing markets by the year 2005. North America will contribute 41 percent of developed market WLL subscribers, followed by Western Europe with 32 percent and Asia-Pacific, led by Japan, with 26 percent [7].

Another important factor in the growth of WLL is the potential conversion of existing residential and business telephone lines as well as new services that can be provided with wireless local loop technology. A good example of this is in rural areas where supplying wired service was previously not cost effective and wireless service can offer high speed Internet access.

Competitive local exchange carriers (CLEC) are competitors to the incumbent local exchange carriers (ILECS) and are likely to use WLL systems to rapidly deploy competing systems. If CLECs do not use wireless systems, they must either pay the existing phone company for access to the local loop (resale) or dig and install their own wire to the local customers. Many countries, that do not have large wired networks such as the United States, are using wireless local loop as their primary phone system.

Figure 4.4 shows a wireless local loop system. In this diagram, a central office switch is connected via a fiberoptic cable to radio transmitters located in a residential neighborhood. Each house that desires to have dialtone service from the WLL service provider has a radio receiver mounted outside with a dialtone converter box. The dialtone converter box changes the radio signal into the dialtone that can be used in standard telephone devices such as answering machines and fax machines. It is also possible for the customer to have one or more wireless (cordless) telephones to use in the house and to use around the residential area where the WLL transmitters are located.

The most basic service offered by wireless local loop (WLL) systems is to provide standard dial tone service known as plain old telephone service (POTS). In addition to the basic services, WLL systems typically offer advanced features such as high-speed data, residential area cordless service and in some cases, video services. To add value to WLL systems, WLL service providers will likely integrate and bundle standard phone service with other services such as cellular, paging, high speed Internet or cable service.

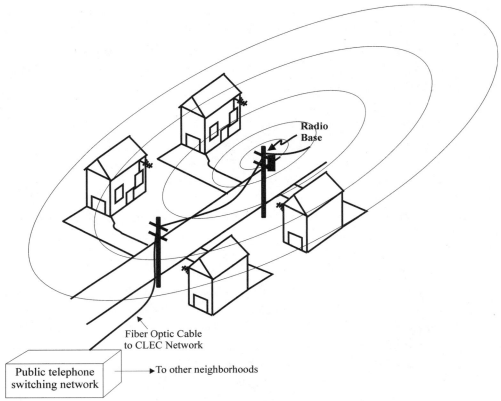

Figure 4.4, Wireless Local Loop (WLL) System

WLL systems can provide for single or multiple-line units that connect to one or more standard telephones. The telephone interface devices may include battery back up for use during power outages. Most wireless local loop (WLL) systems provide for both voice and data services. The available data rates for WLL systems vary from 9.6 kbps to over several hundred kbps. WLL systems can be provided on cellular and PCS, private mobile radio, unlicensed cordless, and proprietary wideband systems that operate in the 3.4 GHz range.

Point-to-Multipoint (PMP) Microwave

Point-to-Multipoint (PMP) systems use hubs to connect several end users to high-speed networks. PMP systems typically use microwave frequencies to provide high speed data services. The wide bandwidths available at microwave frequencies allow the use of data rates that approach fiber links. PMP wireless systems connect a high-speed data user's directional antenna back into the wired world through the use of a hub. The hubs are connected to a backbone network that interconnects hubs to each other and to other networks (such as the public telephone network).

WinStar is one of several companies that provide high-speed point-to-multipoint (PMP) wireless bypass telecommunications services to businesses and residential customers in urban areas in the United States. The WinStar system operates at a frequency of 38 GHz and they have licenses for radio spectrum in more than 125 Metropolitan Statistical Areas (MSAs) throughout the United States [8]. The WinStar system is a combination of high-speed data wireless radios that are inter-connected to a fiber optic network.

The WinStar system is called "Wireless Fiber[SM]." Each 100 MHz Wireless Fiber channel can support transmission capacity of one DS-3 at 45 Mbps. The WinStar's development of multipoint facilities that is planned to begin 1998 should allow a single 100 MHz Wireless Fiber channel to support one OC-3 equivalent of capacity at 155 Mbps.

WinStar identifies strategically located sites to serve as hubs in each of its metropolitan areas. These hub sites are connected via wireless fiber links to customer buildings. Certain characteristics of the 38 GHz frequency, including the effective range of its radio signal and the small amount of dispersion (i.e., scattering) of the radio beam as compared to the more dispersed radio beams produced at lower frequencies, allow for multiple hub sites using the same channel in a licensed area. Because the antennas are highly directional, it is possible to re-use several 38 GHz channels in a single geographic area.

Wireless Local Area Network (WLAN)

Wireless Local Area Networks (WLANs) allow computers and workstations to communicate with each other using radio signals to transfer digital information. Wireless LAN systems may be completely independent or they may be connected to an existing wired LAN as an extension. While adaptable to both inside and outside environments, wireless LAN equipments are especially suited to inside locations such as office buildings, manufacturing floors, hospitals and universities.

Wireless LANs provide all the functionality of wired LANs, but without the physical constraints of the wire itself. Wireless LAN configurations include independent networks, offering peer-to-peer connectivity, and infrastructure networks, supporting fully distributed data communications. Data rates for WLAN systems vary from 20 kbps to 60 Mbps.

Some wireless LANs also allow a personal area network (PAN). A PAN system normally covers the few feet surrounding a user's work area and provides the ability to synchronize computers, transfer files, and gain access to local peripherals.

Other types of wireless network systems include wireless metropolitan-area networks (WMANs) and wireless wide-area networks (WWANs). WMANs are private wireless packet radio networks often used for law-enforcement or utility applications. WWANs are wireless data transmission systems that cover a large geographic area using cellular or public packet radio systems. These wide area systems involve costly infrastructures, provide much lower data rates (regularly below 20 kbps) and often require users to pay for bandwidth on a time or usage basis. In contrast, on-premise wireless LAN equipment requires no usage fees and provides 100 to 1000 times the data transmission rate.

Figure 4.5 shows a typical wireless local area network (WLAN) system. In this diagram, several computers communicate data information with a wireless hub.

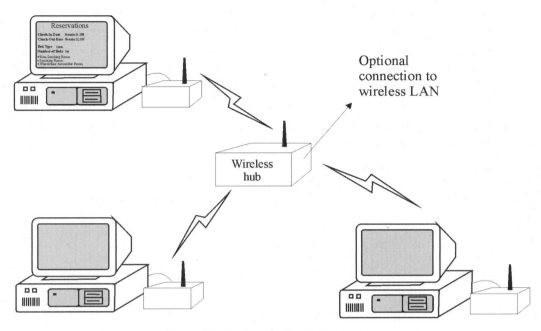

Figure 4.5, Wireless LAN (WLAN) System

The wireless hub receives, buffers and retransmits the information to other computers. Optionally, the hub may be connected to other networks (possibly other wireless network hubs) by wires.

Wireless Local Area Network (WLAN) radio systems use either narrow or wide radio channels or use infrared technology. A narrowband radio system requires coordination of different frequencies to avoid the possibility of interference from other similar frequencies. A wideband radio system uses a radio channel, which is much wider than is necessary to transfer the data information. The extra wide channel is used to spread the signal so it is less susceptible to interference. This is called spread spectrum technology. Infrared technology transfers data in the form of pulses of infrared light. Some short-range infrared systems have shown data transfer rates of 60 Mbps.

The vast majority of wireless LAN products on the market use spread spectrum technology. Spread spectrum products operate primarily within the 900-megahertz and 2.4-gigahertz frequency bands, which do not require FCC licensing. They use limited transmitter power levels (less than one watt) and generally are designed to contain their signaling within 500-800 feet. Some WLAN systems operate in the 5.7 GHz frequency band. In February 1997, the United States FCC approved a plan to make available additional spectrum at 5.15-5.35 GHz and 5.725-5.875 GHz for use by a new category of unlicensed equipment. With these new global frequency bands, this will promote the development of many types of new devices (including WLAN) and improve the ability of manufacturers to compete globally by enabling them to develop products for the world marketplace.

Mobile Wireless Systems

Mobile wireless systems include cellular and PCS (traditional mobile telephones) and 3rd generation wireless (new multimedia wireless). Generally, mobile wireless systems have limited data transmission capability and these systems will not compete with xDSL technologies.

Cellular and PCS Systems

Figure 4.6 shows a basic cellular or PCS system. The cellular network connects mobile radios to each other or the public switched telephone network (PSTN) by using radio towers (base stations) that are connected to a mobile switching center (MSC). The mobile switching center can transfer calls to the PSTN.

When linked together to cover an entire metro area, the radio coverage areas (called cells) form a cellular structure resembling that of a honeycomb. The cellular systems are designed to have overlap at each cell border to enable a "handoff" from one cell to the next. As a customer (called a subscriber) moves through

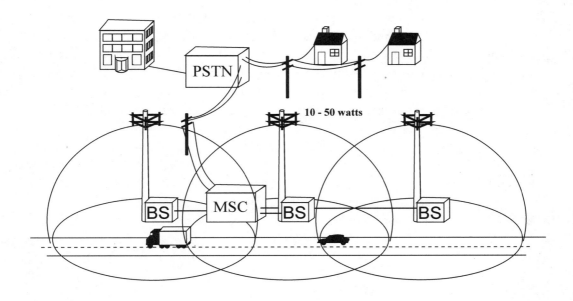

Figure 4.6, Cellular or PCS System

a cellular or a Personal Communications Service (PCS) system, the mobile switching center (MSC) coordinates and transfers calls from one cell to another and maintains call continuity.

Although new technologies have improved the data transmission capability of mobile wireless systems, the maximum data transfer rate for next generation ("3rd generation") mobile systems is approximately 384 kbps in mobile applications and 2 Mbps in local service areas.

3rd generation wireless systems combine multiple types of wireless systems (e.g. cellular and wireless LAN) to provide mobile multimedia (voice, data and video) communications services. The first generation of cellular technology was analog cellular. The second generation of wireless was digital cellular. Digital cellular technology was developed to replace and enhance the capabilities of older incompatible analog cellular standards (e.g. AMPS, TACS, NMT). Although there are only a few digital cellular systems (CDMA, GSM, TDMA), these systems also remain incompatible. The 3rd generation of telecommunications technology proposes a universal mobile telecommunications service (UMTS).

The International Telecommunications Union (ITU) created the concept of the 3rd Generation in 1992 in an effort to consolidate the various 2nd generation standards and combine high speed wired and wireless information services. One of the primary goals of the ITU's UMTS effort is to create one global standard that will be able to allow for a single customer to access fixed or wireless local

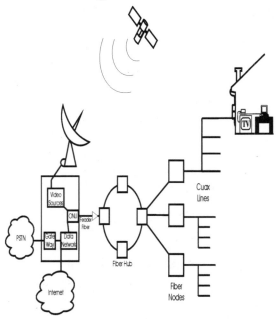

Figure 4.7, Evolution of 3rd Generation Mobile System

loop (WLL) networks. This access would use a common air interface and a Family of Systems Concept (FSC). Such a standard is technically possible, although unlikely.

The technical requirements of 3rd generation systems include backward compatibility for 2nd generation handsets and minimum data transmission rates of 144Kbps for mobile applications (cars, etc.), 384 kpbs for portable (walking) and 2 Mbps for fixed applications.

Figure 4.7 shows the evolution of the wireless systems towards the 3rd Generation wireless system. In this diagram, it can be seen that the vision of the 3rd generation system is the combination of various wired and wireless technologies into a single system.

One of the key requirements for wireless systems is the interconnection of base stations and a single city may have over 1,000 base stations. 3rd generation wireless systems offer much higher bandwidth than their 1st and 2nd generation predecessors. This requires higher speed interconnections between base stations.

Satellite Systems

Satellite systems provide information services to wide geographic coverage areas, typically in one-direction. Satellite systems can provide information services to geographic regions that cannot be served by telephone lines that have xDSL service. For more than 30 years, satellites have been providing voice and data communication service around the globe; however, the cost for equipment and services has been very high.

In 1997, the high cost of satellite equipment and service began to reduce dramatically. New high capacity satellites and digital technology allow for lower cost service and advanced messaging services. Early satellites were analog. After the development of digital satellites, which offer more capacity, several more satellites were put into orbit, followed by the next-generation of low orbiting satellites. These new developments are rapidly bringing the cost of equipment down by over 75%.

Although not commonly known to those in developed nations, more than half the world's population lives more than two hours travel time from the closest telephone! Satellite communications are providing a way to service these remote areas with telephones, news and information feeds.

Satellites orbit in free space, where there is little or no air. In such an environment, there is little to slow the satellites down or wear them out once they are sent into orbit. The useful functional lifetime of a satellite is generally more dependent upon fuel reserves and technological obsolescence than on wear. Satellites are typically classified by the type or height of the orbit they have been placed at around the earth. There are three classes of satellites in orbit today: geosynchronous earth orbit (GEO), medium earth orbit (MEO) and low earth orbit (LEO). GEO satellites are positioned high above the earth (approximately at 22,300 miles) and a single satellite can cover one third of the surface of the earth. MEO satellites are commonly positioned up to 6,000 miles above the earth and a single one can cover several thousand miles. LEO systems are located at approximately 500-1,000 miles above the earth and a single one can cover a thousand miles. There are three basic types of satellite telecommunications services: broadcast international trunking, very small aperture satellite and mobile satellite service. GEO satellites are the only type of satellite that appear stationary (fixed in location) to receivers on earth compared to MEO and LEO satellites that regularly move across the horizon. There are three key portions to satellite systems: satellite section, ground section, and end user equipment.

A satellite is a space vehicle that orbits the earth, and which contains one or more radio transponders that receive and retransmit signals to and from the earth. The size and weight of satellites varies from 1 to over 20 meters in length and 90 lbs (20 kg) to over 8800 lbs (4000 kg). There are several 4000 kg GEO satellites being prepared for launch in 1998. Satellites contain a power supply, position control system, transmitters and receivers (called transponders), and an antenna system.

The power supply for a satellite ordinarily consists of solar panels and a backup battery. The amount of power used by satellites varies from a few hundred watts for small low earth orbit satellites to several thousand watts for large high earth orbit satellites.

Satellites do not automatically stay in their desired location, because of gravity effects from the irregular earth, sun, moon, and other planets. The position of the satellites must be continuously monitored and adjusted (an activity called "station keeping"). The ability to control the stability (wobble) of a satellite is controlled by having a section of the satellite spin at 50 to 100 revolutions per minute. The speed or amount of spin is able to control the angle or relative position of the satellite. The general altitude is normally controlled by a pressurized gas system (generally hydrazine). This means that the life of the satellite system is usually determined by the amount of hydrazine it can carry.

For example, GEO satellites are kept in their correct latitude and longitude by rocket motors that oppose the gravity effects, so they appear to remain stationary with respect to the earth, and ground antennas do not have to track them. The amount of fuel available to do these station-keeping maneuvers generally determines the life of the satellite. Some are allowed to drift north and south and still operate after nearly running out of fuel, because the North-South station keeping uses 100 times as much fuel as the East-West station-keeping. This requires ground antennas to track them in most cases.

The main purpose of satellites is to receive radio signals from earth and retransmit these signals back to earth. Usually, this is accomplished by a transponder. Such a transponder is called "bent-pipe." The transponder receives a signal on one frequency, converts the frequency, amplifies the signal and then sends the signal back to earth. There may be 40 or more transponders on a single satellite, each having a radio channel bandwidth of 80 MHz (or more) Some modern satellites today do on-board digital processing of the signals. These satellites are much more complex and expensive than the simple transponder types described above.

The antenna system of a satellite usually consists of several directional antennas, allowing the satellite to direct its radio energy to specific locations. The radio coverage area provided by a satellite is called its footprint. The footprint of a single satellite can be thousands of miles in diameter, which may be enough to cover an entire continent. Sometimes, however, spot beams are used that cover only a hundred miles or so in order to concentrate signals into a small area and develop stronger signals because of higher antenna gain.

Some systems have multiple satellites (such as MEO and LEO) and these systems often have spare satellites in orbit in the event of an equipment failure. Some GEO systems, such as Intelsat, also orbit spare satellites for both equipment failure and to pick up increased demand.

The ground segment of a communication satellite system contains the gateways that send and receive information signals from the satellite, switching or routing facilities and a satellite control center. Gateways provide access to the space segment and interface to public and private data networks.

End user equipment (typically called subscriber units) consists of an antenna, radio receiver, transmitter (only for two-way systems) and an interface converter (such as a video or audio interface) depending on the application.

There are various types of subscriber units, some of which are intended for general use, and some of which are designed to support specific applications. Subscriber units that are used for mobile or portable voice communication may be capable of several services such as voice, messaging and data. They usually appear very similar to a cellular telephone. Other devices, such as digital television receivers or meter reading devices, are designed specifically for their application.

In 1999, Direct Broadcast Satellite (DBS) offered the most widely available high-bandwidth Internet access technology in a geographic area. DBS service requires a small satellite receiver antenna (a "dish"). DBS service providers can deliver download speeds up to 350 kbps.

Very Small Aperture Terminals (VSATs) are ground-based satellite end user equipment assemblies that offer two-way data access to satellite systems. Typical uses for VSAT networks are the broad distribution of data to many receivers. An example of this service is the distribution of price or inventory levels to a chain of stores throughout a large geographic area. They are also used for each store to send inventory and sales information back to headquarters. An important feature of a VSAT service is ease of deployment; installation takes approximately 2 hours.

Fiber Systems

Optical fiber cable provides for tremendous bandwidth transmission without many of the signal distortion challenges of copper wire. A single strand of fiber can transport over 1 Gbps of data and hundreds of fibers may exist in a single cable.

Although copper wire is still the most popular horizontal cabling choice in the end loop, most high-speed interconnection lines use fiber. The most common reasons for the user of fiber include immunity to electrical interference, constant error-free transmission, long distance between repeaters, higher data transmission rates, ease of handling (it is much lighter than copper) and lower overall maintenance costs for high speed circuits. All of these reasons provide tremendous value for fiber. However, probably the primary reason to use fiber systems is lower cost for communications circuit.

There are three basic types of fiber systems used to service residential customers; fiber to the neighborhood (FTTN), fiber to the curb (FTTC) and fiber to the home (FTTH). FTTN brings a fiber connection to a small geographic area that allows service to be provided to 100 or more houses. Fiber to the curb brings the fiber close to a small group of homes or businesses. Fiber to the home extends the fiber directly into the home or business. FTTH is sometimes also called fiber to the basement (FTTB).

When fiber optic connections are used, the fiber communication line terminates into an optical network unit (ONU). ONU's are used to multiplex and de-multiplex signals to and from a fiber transmission line. An ONU terminates an optical fiber line and converts the signal into a format suitable for distribution to a customer's equipment. When used for residential use, a single ONU can serve 128 to 500 dwellings.

FTTN and FTTC systems are the most common as fiber is used to bring several circuits to residential areas. FTTH systems have been tested with disappointing results as the cost to install FTTH is high and the usage requirement for residential customers (revenue potential per line) is low.

There are two primary types of fiber optic distribution systems; active and passive optical systems. Active systems multiplex multiple channels in time sequence on a single optical carrier signal. When the optical signal is received, it is divided using digital signal processing techniques. Passive systems use different optical wavelengths for each channel. This is called wave division multiplexing (WDM). WDM systems use passive optical filters (no processing or power) to separate each optical channel to its designated receiving device.

There have been many trials of optical systems to the home. Many new high bandwidth applications (such as video on demand) have been discovered as a result of these efforts. However, the installation cost of fiber systems typically prohibits their use in homes. At the beginning of the 21st century, optical were primarily used for interconnection of telephone switches and to delivery groups of circuits to business centers (e.g. office buildings).

Figure 4.8 shows a telephone network that uses and optical carrier system. This diagram shows several types of fiber systems. In the simplest example, a fiber optic line extends from a central office and terminates in an ONU in the center

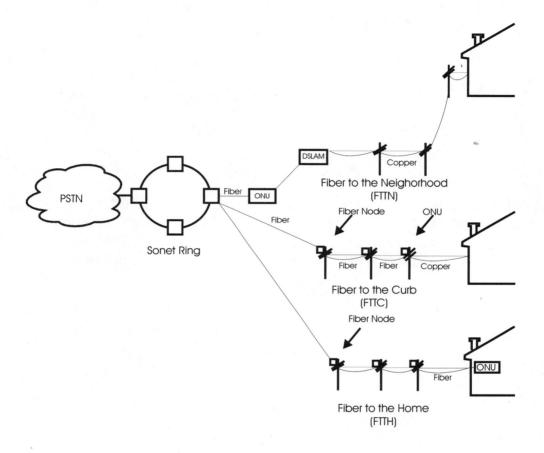

Figure 4.8, Optical Carrier Systems

of a neighborhood. The ONU provides the high bandwidth necessary to connect several DSLAMs that provide ADSL to customers within the neighborhood. As the need for increased bandwidth increases (such as a higher demand for Video on Demand service), a fiber ring is extended around the neighborhood and ONUs (fiber nodes) are installed on telephone poles that allow VDSL technology to be connected to homes within 1000 feet of the VDSL capable DSLAM. Finally, as end users continue to increase the demand for higher bandwidth than even VDSL can provide, fiber can be installed into the home and an ONU is installed within the building, typically in the basement.

References:

1. Globalstar 10K report, 31 March 1998, US Securities and Exchange Commission.
2. 10K report, 31 March 1998, US Securities and Exchange Commission
3. ibid.
4. 2wire.com web site, 18 April 2000.
5. ibid.
6. Interview, industry expert, Broadband Cable 2000 conference, 8 May 2000.
7. "Wireless Resale Market Report," Multimedia Publishing Corporation, Houston, 1997, pg. 301.
8. 10K report, 31 March 1998, US Securities and Exchange Commission.

Chapter 5
End User DSL Equipment

Consumer end user DSL equipment modems, splitters, set top boxes, gateways, network interconnection equipment. When used for businesses, end user equipment may include standard line adaptation interfaces and digital telephone interface devices.

Consumer premises equipment (CPE) is the translator between your computer (or data communications device) and the xDSL network (typically the connection to the Internet). Your CPE performs to key functions; data format adapting and protocol conversions. First, your DSL CPE must convert the digital format from your computer into a format that is suitable for transmission to the DSL network. This involves voltage level conversion, timing and modulation. The second step for your DSL CPE is to adapt the protocols. This includes message structure and communications channel control. This may involve multiple protocols such as asynchronous transfer mode (ATM) and Internet protocol (IP).

If you have more than one computer at home, they can all be connected to one DSL modem using a home network. One option is to buy an Ethernet hub and connect all your computers to it, much like a small-office local area network (LAN). You can then connect the hub to your DSL modem and all the computers can access the DSL connection. There are some DSL modems that include an Ethernet hub. One disadvantage of this approach is that you would need to install special wiring throughout your home to connect the computers. Another option is to use one computer as a "gateway" to other computers in the home via home networking technology.

There are three types of home networking that don't require any new wiring in the house: powerline, phone line, and wireless. Powerline networking technology uses the electrical wiring and outlets of your home to create a network. Phoneline networking does the same thing using the telephone wiring and outlets—and it does not interfere with phone calls on the same wires. Wireless technology accomplishes the task using two-way radio waves transmitted through the house. Overall, using a PC as a gateway has the disadvantages of: requiring some technical expertise, requiring the gateway PC to be turned on for other PCs and networked devices to use the Internet connection, and lacking reliability as PCs often crash or lock up.

xDSL Modems

xDSL modems (MOdulator/DEModulator) convert digital signals to and from complex analog signals (that represent the digital information) for transmission and reception over conventional analog telephone wires. xDSL modems that can deliver multi-megabit data services. There are two basic types of xDSL modems; computer (internal) modems and external. Internal modems are cards that are installed inside a personal computer via a plug-in card. External modems can be connected to a computer via a parallel connection, USB port, or Ethernet jack.

Because DSL standards are subject to a manufacturer's interpretation, not all xDSL modems can communicate with xDSL systems of the same type. It is important to make sure that the DSL modem (sometimes referred to as a "terminal adapter") works with the selected DSL service provider's equipment. The DSL modem should be included in the package with the high-bandwidth service sold to you by the DSL provider. This will probably change in the future as more manufacturers produce equipment based on universal standards. If you buy a modem off the shelf, ask if it is compatible before you order service from a DSL provider.

DSL CPE is commonly called a DSL modem. Although all DSL CPE devices contain a modem, there are several different DSL CPE device configurations that include more than a modem. The three most common methods of connecting a DSL CPE device to a computer include Ethernet, universal serial bus (USB) and peripheral component interconnect (PCI).

There are two basic types of DSL modems; routers and bridges. A DSL router is capable of translating and switching communications channels. A DSL bridge only adapts existing communications channel to your computer or data device.

DSL modems and other CPE devices contain firmware that controls their operation. Firmware is software program instructions that are stored in a hardware device that performs data manipulation (e.g. device operation) and signal processing (e.g. signal modulation and filtering) functions. Firmware is stored in memory chips that may or may not be changeable after the product is manufactured. In some cases, firmware may be upgraded after the product is produced to allow performance improvements or to fix operational difficulties. It is preferable to obtain DSL CPE equipment that has the ability to upgrade firmware after the product is purchased. This may allow for repair of operational difficulties or enhancements to product performance.

Ethernet Adapter

For the first generation of xDSL connections, Ethernet adapters were one of the most popular methods of connecting computers to DSL networks. Ethernet connections come in two common forms; 10 BaseT (10 Mbps) and 100 BaseT (100

Mbps). Although Ethernet adapters were not standard equipment on most personal computers in 2000, Ethernet equipment was readily available in 2000 from a variety of manufacturers.

An Ethernet hub is a communication device that connects several devices in a network (often data communication devices). A hub, generally, is a simple device that distributes data messages to multiple receivers. However, hubs can include switching functional and multi-point routing connection and other advanced system control functions.

Some Ethernet DSL modems include a built-in-hub. This allows more than one device to be connected to the DSL connection. These devices can be 10 BaseT, 100 BaseT or some CPE hubs can automatically switch between 10 BaseT and 100 BaseT. Automatic switching between 10 BaseT and 100 BaseT can be a significant benefit when connecting various types of accessories to the Ethernet. Accessories such as printers that may only have a 10 BaseT connection while desktop computers may have 100 BaseT. Laptops may only have an auto switching 10 BaseT/100 BaseT PCMCIA network adapter card.

External USB Modem

Most computers that were manufactured in 2000 included a universal serial bus (USB) connector. USB is a standard communication bus that can transfer data at speeds up to 12 Mbps. The USB data bus can also connect several devices to the same bus using a low cost hub device. USB lines can only extend for a few feet from the computer. Figure 5.1 shows an external USB modem adapter. This device converts ADSL signal into standard USB format.

Figure 5.1, External USB ADSL Adapter

Source: Efficient Networks

Computer Modem (PCI) Card

The computer modem card is also called a network termination interface (NTI). An NTI can be a variety of adapter cards other than an xDSL modem. One of the more common interfaces to a personal computer is a peripheral component Interconnect (PCI) bus. A PCI bus is a standard data communication connection that allows accessory cards to be installed into a personal computer. The PCI specification defines both the electrical and physical (connector) requirements and was introduced to the marketplace by Intel. The PCI bus allows up to 10 PCI-compliant expansion cards in a PC. The PCI standard has replaced the previous industry standard architecture (ISA) bus.

Typically, when a xDSL computer modem card is installed in a personal computer, software configuration is necessary. For later generations of Windows products, the operating system will automatically sense the installation of a modem.

Figure 5.2 shows an ADSL modem card that installs into the PCI socket in a personal computer. This card connects to the peripheral component interconnect (PCI) card. Several standard PCI interface adapters are usually located inside a personal computer.

PCMCIA Card

The personal computer memory card international association (PCMCIA) standard interface allows NIC cards and other data communication devices to be installed in portable computers. In most applications, only an Ethernet or Wireless Ethernet PCMCIA card is used in portable computer to connect to a DSL modem that has an Ethernet port.

Figure 5.2, ADSL Computer Modem Card

Source: Xpeed

Splitter

An xDSL splitter separates two signals; typically analog telephone (POTS) and digital (e.g. ADSL digital channel). There are two types of splitters: passive and active. A passive splitter simply separates low frequency and high frequency components of a xDSL signal. An active splitter includes electronics that can provide some test capability and signal amplification.

Set Top Box

A set top box is an electronic device that adapts a communications medium to a format that is accessible by the end user. Set top boxes are commonly located in a customer's home to allow the reception of video signals on a television or computer. A set top box is commonly used to convert digital video (e.g. MPEG2) into standard NTSC video that is used for televisions.

Residential Gateway

Residential gateways are a simple bridge that is located between a DSL network and a home LAN. It performs as a DSL modem and a home networking hub for multiple PCs. The main advantages of a residential gateway will be their ease of use and reliability. More intelligent residential gateways, called multi-service residential gateways, will provide additional capabilities, such as enhanced telephone features and entertainment services. 2Wire's Home Portal is a multi-service residential gateway. Figure 5.3 shows a residential gateway produced by 2Wire.

Figure 5.3, Residential Gateway

Source: 2Wire

Premises Distribution Network (PDN)

A premises distribution network (PDN) is a network that is located at a customer's facility. A PDN is used to connect terminals (computers) to other networks and each other. For xDSL systems installed in 2000, the most common type of PDN was EtherNet. Other types of PDN networks include ATM 25, Universal Serial Bus (USB), Home Packet Data Network (HomePDN) and FireWire (IEEE-1394).

Ethernet

Ethernet is a packet-switching transmission protocol that is primarily used in local area networks (LANs). Ethernet exists in various forms including 10 BaseT and 100 BaseT. Ethernet can be provided on coaxial cable, twisted pair or wireless. When Ethernet systems use wired connections, the data transmission rate is 10 Mbps or 100 Mbps. 100 BaseT systems are also called "Fast Ethernet." When Ethernet systems use wireless links, the data transmission rates are limited to 1 Mbps to 11 Mbps. Wireless Ethernet systems typically use the ISM frequency band in the 2.4 GHz range. Wireless networks typically operate up to a distance of 300 feet (100 meters).

Ethernet systems commonly use a hub to distribute data to multiple devices. A hub is generally a simple device that buffers (amplifies) and distributes data messages to multiple receivers. However, hubs can include switching functional and multi-point routing connection and other advanced system control functions.

ATM 25

ATM 25 is a 25 Mbps version of asynchronous transfer mode (ATM). The ATM 25 standard was developed primarily for corporate networks. However, the QoS advantages of ATM and customer needs for switched services for digital video and Internet access has stimulated interest in ATM for the DSL industry.

Phoneline Networking

In the late 1990's, the home phoneline network alliance (HomePNA) developed a specification that allows home computers and data devices (such as network printers) to interconnect via standard telephone lines. In the first generation of phoneline networking, data rates of 1 Mbps were achieved. Data transmission rates of 10 Mbps have been demonstrated. The Phoneline Network uses high frequency signals that do not interfere with standard telephone service. To connect a phoneline network to a DSL connection, a phoneline bridge can be used.

Universal Serial Bus (USB)

Universal serial bus (USB) is a short distance data communication interface (typically, only a few meters) that is installed on most personal computers. The USB was designed to replace the older UART data communications port. USB ports permit data transmission speeds up to 12 Mbps and up to 10 devices can share a single USB port.

FireWire

FireWire is a short distance data communications interface based on industry standard IEEE-1394. FireWire can transmit at speeds up to 400 Mbps.

Figure 5.4 shows the various configurations of the premises distribution network. This diagram shows that an ADSL line enters the customer's premises in a variety of ways. In the first configuration, the modem is connected to an Ethernet hub. This allows multiple computers to share the same high-speed data line. In another configuration, the xDSL modem is converted to ATM 25. The ATM 25 system divides up the data channel for multiple types of usage including digital video. Alternatively, a bridge connects the xDSL modem to a HomePDN system. The home PDN system routes the data to PCs and other HomePDN compatible devices through the existing wiring in the home. A common single user configuration is the USB data connection. An external xDSL modem is connected to the USB port on a computer. FireWire is very similar to USB except it allows the connection of more devices and provides up to 400 Mbps of data transfer rate.

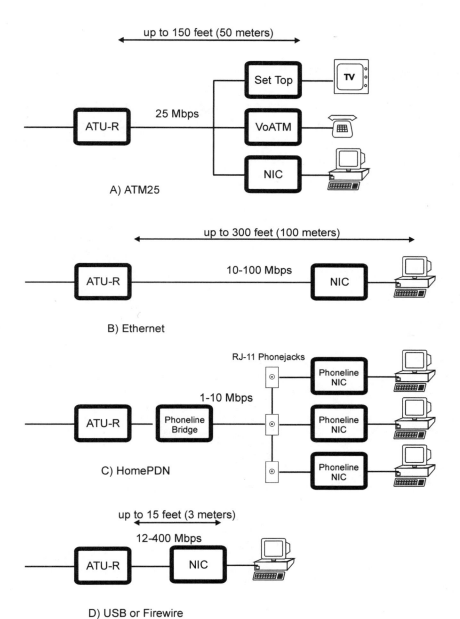

up to 150 feet (50 meters)

Set Top — TV

ATU-R — 25 Mbps — VoATM

NIC

A) ATM25

up to 300 feet (100 meters)

ATU-R — 10-100 Mbps — NIC

B) Ethernet

RJ-11 Phonejacks

Phoneline NIC

ATU-R — Phoneline Bridge — 1-10 Mbps — Phoneline NIC

Phoneline NIC

C) HomePDN

up to 15 feet (3 meters)

12-400 Mbps

ATU-R — NIC

D) USB or Firewire

Figure 5.4, Premises Distribution Networks (PDNs)

Proxy Server

A proxy server is a computing device (typically a server or personal computer) that interfaces between data processing devices (e.g. computers) that are connected to a local network and an external network (e.g. the Internet). A proxy server has two network interfaces. One interface communicates with the local area network (e.g. Ethernet) and the other network interface (e.g. DSL modem) communicates with another data network.

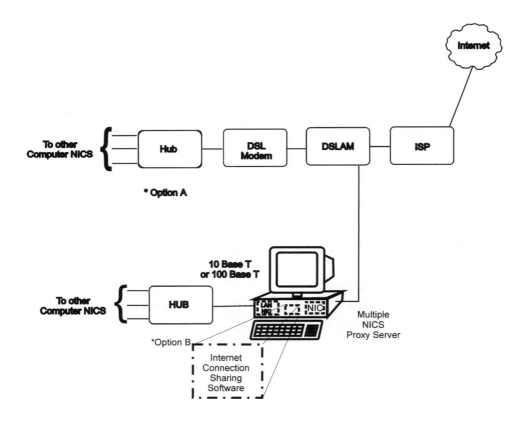

Figure 5.5, Proxy Server

Proxy servers add a significant amount of security protection to LANs that are connected to public data networks. Most proxy servers support network address translation (NAT) and dynamic host configuration protocol (DHCP). A proxy server requires proxy server software to operate. Examples of proxy server software include WinGate, Microsoft Proxy Sever and Win Proxy.

Figure 5.5 shows a end user's system that takes advantage of a proxy server. This diagram shows that an Ethernet LAN interfaces to the Internet through a proxy server. The proxy server in this diagram accepts messages from private IP addresses (from the local LAN), translates these messages to an public IP address which are sent to the Internet via the DSL modem NIC. When the Internet responds to the translated messages, the proxy server routes (addresses) the received message to the appropriate computer.

Digital Service Unit (DSU)/Channel Service Unit (CSU)

Digital Service Unit (DSU) adapts a standard communications circuit (e.g. T1) to a different medium (e.g. twisted pair HDSL). Digital service units are commonly used in commercial applications to convert data communications to standard transmission line formats (such as T1 or E1). HDSL2 DSUs can provide the equivalent of a T1 or E1 circuit over copper wire pairs.

Integrated Access Device (IAD)

An integrated access device (IAD) converts multiple types of input signals into a common communications format. IADs are commonly used in PBX systems to integrate different types of telephone devices (e.g. analog phone, digital phone and fax) onto a common digital medium (e.g. T1 or E1 line).

Software

It is necessary to configure your operating software to work with the DSL modem. Fortunately, getting DSL to work with Windows 95/98, Windows NT and MAC is relatively easy and generally already supported in the operating system software. However, some DSL modem devices will require some software installation.

When installing a DSL modem, the computer must be configured via a network interface card (NIC) to work with TCP/IP. For Windows products, dial up networking (DUN) may be used. DUN is a software portion of Microsoft Windows 95,98,NT, and 2000 that allows the user to connect the computer to a data network (such as the Internet). DUN is actually a process of establishing and maintaining a communications session. However, DUN is sometimes used for establishing connections on "always-on" circuits (such as DSL).

There are many types of optional software that may be used with DSL connections. One of the more popular software applications is software that allows telephone calls to be placed over the Internet.

Security

When connecting a computer to the Internet or other public data network, security issues should be considered. This is especially important when the connection to the public data network is constant ("always on"). There are a variety of low cost security options available for DSL CPE devices and systems. You want to limit the ability for outside users to connect to your computer and restrict the types of operations that can be performed from outside connections when they are necessary.

Network Address Translation (NAT)

One of the best security measures is network address translation (NAT). NAT converts your private Internet address into an Internet address that can be transmitted on the Internet. This effectively does not allow unsolicited access from

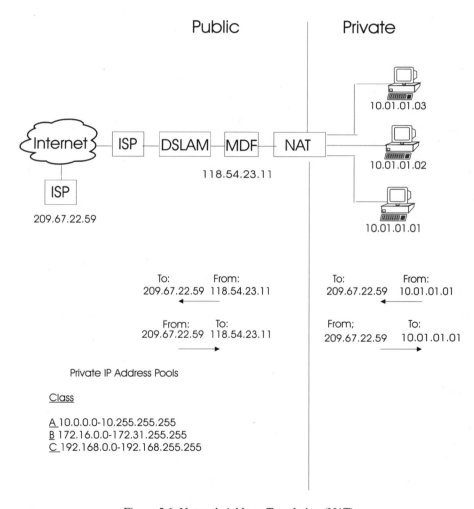

Figure 5.6, Network Address Translation (NAT)

outside computers to your computer. Unfortunately, this can complicate video-conferencing where a direct connection is necessary for automatic videoconferencing setup.

Figure 5.6 shows the basic operation of NAT. This diagram shows that when the NAT receives a message with a desired pubic IP address (209.67.22.59) originating from a local computer with a private IP address of 10.01.01.01. The NAT translates this originating address to a public IP originating address. The he NAT then initiates as session with the Internet server (web site) using the network's public IP address 118.54.23.11 as the originating address. The Internet server receives the request for information and responds with data messages address to the NAT's public IP address. When the NAT receives these data messages for that particular communications session, they are translated to the local (private) IP address 10.01.01.01 and forwarded to the originating computer. If messages are received to the NAT's public IP address that are not part of a communications session that it knows about, the NAT will not route the messages to computers connected to the LAN.

Dynamic Host Configuration Protocol (DHCP)

Dynamic host configuration protocol (DHCP) is a process that dynamically assigns an Internet Protocol (IP) address from a server to clients within a network on an as needed basis. The IP addresses are owned or controlled by the server and are stored in a pool of available addresses. When the DHCP server senses a client needs an IP address (e.g. when a computer boots up in a network), it assigned one of the IP addresses available in the pool.

Figure 5.7 shows the basic operation of DHCP. This diagram shows that when a computer is powered on (booted), it requests connection to the network. When the DHCP server receives the request, it reviews its list of available IP addresses and assigns a temporary address to the requesting computer.

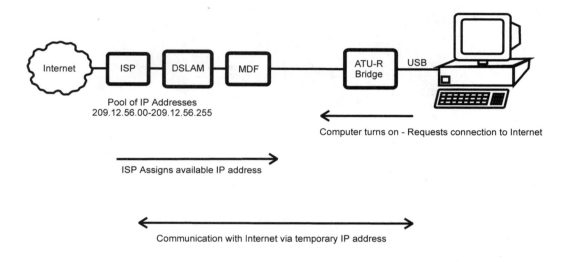

Figure 5.7, Dynamic Host Configuration Protocol (DHCP)

Firewall

A firewall is a device or software program that runs on your computer that provides protection from Internet or data network intruders. Firewalls can restrict access types and may monitor for advanced security threats by analyzing certain types of data communication activities. Although firewalls are important for protecting data that is connected to pubic networks, they can be complicated to setup, can cause problems with desired communications and generally can slow down the transfer of data communication.

Figure 5.8, SOHO Firewall

Source: SonicWALL

Figure 5.8 shows a picture of a firewall that is used for small office and home office (SOHO) applications. The SonicWALL firewall can protect up to 10 computers. This device connects between the LAN and the DSL modem.

IPSec

There are Internet Protocol Security (IPSec) features in IP version 6 (IPv6). Unfortunately, the implementation of IPv6 is likely to achieve significant market share until 2005.

Security Tips

An ideal firewall is called a "brick wall" firewall. Firewalls add delay time to data processing and highly secure firewalls can costly and are complex to setup. There are some basic steps you can take to increase security without having to install very expensive firewalls.

First, disconnect your computer from the Internet when you are not browsing the Internet. Unauthorized users ("hackers") cannot access your computer when it is disconnected.

Disable your mail preview option. This prevents you from inadvertently activating scripts and viruses that may be embedded in email messages. If you don't know where the email came from and are suspicious of its contents, email the original sender to confirm the validity of the email prior to opening the email.

Get dynamic with your email address. Static IP addresses allow hackers to find you every time you log on. If your IP address keeps changing, it will be harder for a hacker to access your computer for extended periods of time.

There are low cost software packages and hardware equipment that offer a moderate level of increased security. They may not stop all hackers, but they will likely stop most of them.

Installing End User DSL Equipment

Installing end user equipment involves choosing the equipment configuration options and possible line wiring changes.

Equipment Configuration Options

The line configuration options will depend on the available equipment (compatible equipment) with the DSL service provider's network and on the existing wiring and ability to add wiring at a customer's location. In general, it is better to have short cable runs and a minimum number of cable splices at the customer's location.

For some xDSL technologies (e.g. ADSL), a splitter may be required to allow the simultaneous use of existing analog telephone equipment and data communication equipment. The location of this splitter is a key configuration option. The splitter can be located near the network termination (NT) at the building entrance or in a basement. It can also be located inside a personal computer or eliminated in some situations through the use of microfilters. The use of microfilters often allows the customer to install their own DSL equipment. This may eliminate the need for an installer to visit with the customer (eliminate the "truck roll").

Figure 5.9 shows the most popular types of DSL installation options. In example A, an ADSL modem and splitter is installed near the network termination (NT) point. The ADSL splitter provides a plain old telephone service (POTS) output via an RJ-11 modular connector. The ADSL modem provides an Ethernet

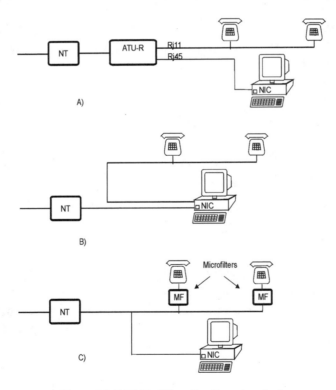

Figure 5.9, DSL End User Configuration Options

output via an RJ-45 modular connector. Because Ethernet cable is not readily available in most homes or business, a computer cable must be installed between the ATU-R and the computer. In option B, the ADSL modem has the splitter located within a network interface card (NIC). For this option, a new phone line must be located between the NT and the DSL modem. The NIC contains a RJ-11 modular plug that connects POTS service telephones. This option requires that a POTS line be run from the computer back to other analog phones. The benefit of option B is a short cable run between the NT and the computer. For option C, ADSL-Lite (also called G.lite or CDSL) is used and no splitter is required. The ADSL-Lite modem or NIC is installed in the personal computer. To minimize interference to and from other POTS telephones, microfilters (audio bandpass filters) are installed between the POTS telephones and the wall jacks.

Line Installation

The typical line installation for xDSL equipment involves converting existing lines to xDSL capability or adding new wire to a remote location in the building (such as a room used for family activities or home office).

Converting existing cabling systems to allow for xDSL service may involve removing of excess wire from connection points, reducing the number of splices and repairing poor splices. When installing new cable for xDSL service, unshielded twisted pair (UTP) wire is commonly used in most residential and small business locations. One of the most popular UTP configurations is 4 wires each separately insulated and all covered in one outer jacket. These wires are either 26 AWG or 24 AWG (usually 24) and are all running parallel to each other.

Originally (many years ago), telephones required 4 wires to operate. Today, most analog and digital telephones only require one pair of wires (2 conductors). This allows many installations to use the same existing wire for two separate phone numbers. The standard color selection for telephone wires are most always red, green, black and yellow. The normal connection uses red and green for the first phone line and black and yellow for the second line. Many times this will result in faint or sometimes significant line noise that is referred to as crosstalk.

Crosstalk is noticed when both lines are in use at the same time and a faint echo of another conversation can be heard in the background. Although crosstalk can be tolerated in most analog phone systems, this usually causes interference that reduces the data transmission rate in DSL systems.

Although the use of UTP for each pair of communication wires reduces the amount of crosstalk (distortion due to signal leakage), the quality of wire construction, insulation and termination connectors also can provide a significant reduction in the amount of interference. Installation of the cable can also have a significant effect on the performance of DSL transmission. High frequency signals can be easily disrupted if the cable is installed with hard turns, when too much tension on a cable pull is applied, allowing the wire to kink, and untwisting pairs over 1/2 inch at termination, etc. Generally, telephone installation professionals or trained installers should be used to upgrade home wiring for DSL service.

There are several categories of twisted pair communication wires that relate to the line speed at which the wire is capable of passing data. This is related to the manufacturing specifications of the wire and is related to the quality of the wire, individual insulation characteristics, and number of twists per inch. If the wiring is to be used for digital data applications it should at least meet Category-3. Category-5 is required for most industrial and office applications where there is a heavy reliance on data traffic. For a complete installation to meet Category-5 or higher much care must be taken through out the installation process.

In residential and small office applications, anything over Category-3 is usually adequate as related to cost and realistic ability to transfer digital signals. In most residential applications the runs are short and thus the difference in line speed would not be noticeable. It is still recommended to use Category-5 wiring for any application where digital service will occur. The difference would be the expense of a specialist to assure a quality installation in a residence. Using an unqualified homeowner, handyman, cable/phone guy, etc. who may improperly install cable or attach a telephone jack or accessory may reduce the data transmission performance of the cable. Although the data transmission may still occur, the data transmission rate may be lower than necessary due to signal distortion.

The connectors also can have an effect on data transmission performance. The common types of telephone connection are RJ-11 or RJ-45 plugs. Companies such as Innovative Connectors ("InCon") offer quality cabling and universal connectors that permit different types of connector plugs to telephone and data devices.

Figure 5.11 shows a Multi-Function Communication Faceplate that supports 3 wire types through one hole in the wall. Power, Coaxial Cable and twisted pair communication wire (phone/digital). The InCon Tri-Plate attaches directly to a standard decorator style A/C receptacle that is mounted conventionally in a standard single size electrical outlet box. The InCon cable that is used for these universal connectors consists of one RG-6 Coaxial Cable that is thinly webbed to a 4 pair Category-5 bundle of Twisted Pair communication wire. For a Local Area Network (LAN) a cable is made up with the LAN (RJ-45) computer jack at one end and 2 phone type (RJ-11) jacks at the other. This leaves the other 2 RJ-11

Figure 5.11, Universal Communications Connector
Source: Innovative Connectors ("InCon")

jacks and their respective twisted pairs for use for phone or other devices. Additionally, since the "special" wiring configuration is in an external jumper it is easily changed as technology shifts. Typical RJ-45 jacks wired in the wall are wired with all 4 twisted pairs when only 2 are used in most available networks.

Chapter 6

DSL Networks

A DSL network is composed of several key parts; this includes a local access line provider, DSL access provider, backbone network aggregator, ISP provider and other media providers. DSL services can be provided by a single service provider or may result from the combination of processes from different service providers. The communication network can be divided into several parts; local access lines (copper), voice communications network (PSTN), high-speed access network (xDSL), aggregator (interconnection), Internet service provider (ISP) and content provider (media source). These network parts and the service providers who operate them, must interact to provide most DSL services.

Network Overview

The physical parts of a DSL network include a subscriber access device, network access lines and adapters and interconnected to service provider equipment. There are many configuration options for a xDSL network. They vary from a simple end user's modem bridge that connects a single end-user's computer to the DSL network to complex multi-channel ATM systems that connect routers and set top boxes.

Figure 6.1 shows the functional parts of DSL network. This diagram shows that end user equipment adapts, or converts analog and digital signals to a high-speed xDSL transmission signal via a DSL modem (an ATU-R for an ADSL system). The copper wire carries this complex xDSL signal to a DSL modem at that connects to the central office (an ATU-C for an ADSL system) where it is converted back to its analog and digital components. The analog POTS portion of the signal (if any) is routed to the central office switching system. The high-speed digital portion is routed to a digital subscriber line access module (DSLAM). The DSLAM combines (concentrates) the signals from several ATU-Cs and converts and routes the signals to the appropriate service provider network.

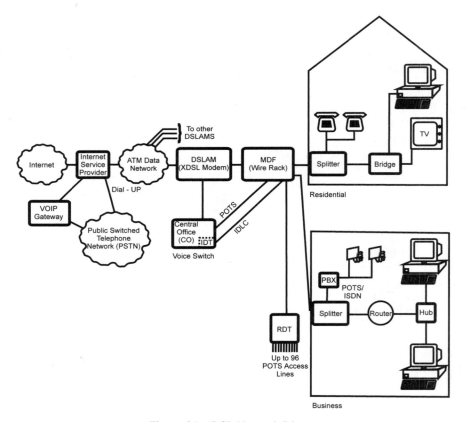

Figure 6.1, xDSL Network Diagram

DSL Modem - Remote

A DSL modem is an advanced modem that provides for high-speed data (up to multi-megabit data rates) over unshielded twisted pair (UTP) of copper wires. The DSL modem is usually located at the customer's home or business office. When used for asynchronous digital subscriber line (ADSL), the modem is called ATU-R.

The ATU-R can be in various configurations including and internal computer modem (PCI bus), external modem that connects to the Universal Serial Bus (USB), or a bridge device that converts ADSL signals to a 10BastT or 100BaseT Ethernet form.

xDSL Splitter

The xDSL Splitter is a device or component that divides an xDSL signal into separate voice and data outputs. A xDSL splitter is typically used for ADSL and VDSL systems.

The xDSL splitter separates the existing telephone signal from the high-speed data signal. In the United States, the standard telephone signal (POTS) frequency band extends up to 8 kHz. In Europe, standard telephone signals include additional high frequency components that extend up to 12 kHz. When the xDSL splitter is used to allow ISDN telephone signals, the frequency band must extend up to 80 kHz (120 kHz for ISDN in Germany). Figure 6.2 shows the differences in frequency bands for the ADSL splitter.

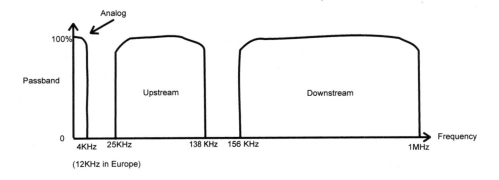

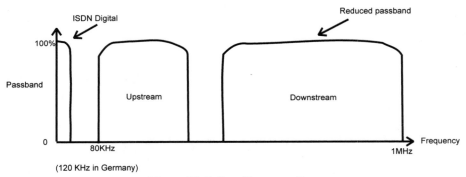

Figure 6.2, Splitter Frequency Range

Network Termination

A network termination (NT) is the final end point in a network that is usually owned by the network service provider. After the network termination point, additional cable and equipment that is located on the premises is commonly owned by the end-customer. This equipment is called customer premises equipment (CPE).

The network termination (NT) can be an active device or passive device. When it is an active device, the NT typically has standard communications parameters such as protocols, timing and voltages to allow specific types of equipment to correctly communicate with the network. When it is a passive device, it just isolates the network from the customer's telephone equipment.

Transmission Line

The transmission line for xDSL systems is typically unshielded twisted pair (UTP). UTP consists of pairs of copper wires twisted around each other and covered by plastic insulation.

The twisting of the copper wire pair reduces the effects of interference as each wire receives approximately the same level of interference (balanced) thereby effectively canceling the interference. UTP is by far the most popular cabling used for local access lines and computer LANs (such as 10BaseT and 100BaseT).

Copper Cross Connect

A copper cross connect system allows access lines (copper lines) to be connected to several different xDSL modems. Copper cross connects are optional devices (not all DSL systems use copper cross connects). There are two key reasons to use a copper cross connect system. The first reason is to allow a copper wire access line to be connected to different digital subscriber line modems. This could be because the customer may upgrade to a new type of modem (e.g. ADSL to VDSL) or if a specific xDSL modem fails, a spare DSL modem could be connected to the customer's line.

The second reason to use a copper cross connect system is to allow an access line to be connected to a DSL modem only when a connection is required. This would allow a DSL service provider to install lesser number of modems in a system than they have customers for.

DSL Modem – Central Office

A central office DSL modem communicates is essentially a mirror image of a DSL modem that is located at the customer premises. The DSL modem on the system network side may be located inside the central office or may be co-located at a remote digital terminal (RDT) site.

The DSL modem at the central office (or remote access node) converts the complex DSL signal into a digital data signal that can be routed through the backbone network. When used with ADSL technology, the DSL modem is called an ATU-C.

Figure 6.3 shows a block diagram of an ATU-C modem. This diagram shows that the key functional parts of the ADSL network modem consists of a digital line interface, digital signal processing section for channel coding, a modulator, line driver and hybrid assembly. The interface converts the data format that is supplied by the DSLAM assembly (e.g. Ethernet or ATM) and converts into a format that can be manipulated by the modems digital signal processing section. The coded signal is modulated and then supplied to a line driver. The line dri-

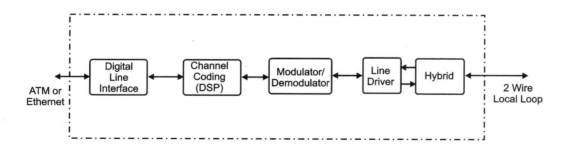

Figure 6.3, Block Diagram of an ATU-C

ver supplies and receives the DSL signal via the hybrid assembly. The hybrid assembly combines and separates transmit and receive signals from the 2-wire copper access line to 4 wires.

Digital Subscriber Line Access Module (DSLAM)

The digital subscriber line access module (DSLAM) usually holds several digital subscriber line (DSL) modems that communicate between a telephone network and an end customer's DSL modem via a copper wire access line. The DSLAM concentrates (also known as a "Concentrator") multiple digital access lines onto a backbone network for distribution to other data networks (e.g. Internet). Figure 6.4 shows DSLAM concentrator.

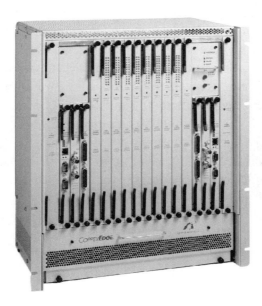

Figure 6.4, The Copper Edge 200 DSL Concentrator

Source: Copper Mountain

The DSLAM is usually mounted in the central office. Because the amount of space in a central office (CO) is limited, DSLAMs may be located at a nearby location. This is not an ideal solution as this adds cable length and splice points to the access line. Additional cable length may reduce the data transmission rate.

Network Routing and Switching Equipment

The network routing and switching equipment provides communications paths between the end user and the services they desire to use (e.g. Internet). The two basic types of switching equipment that is used are routers and ATM switches.

Router

A router is a device that directs (routes) data from one path to another in a network. Routers base their switching information on one or more information parameters of the data messages. These parameters may include availability of a transmission path or communications channel, destination address contained within a packet, maximum allowable amount of transmission delay a packet can accept along with other key parameters. Routers that connect data paths between different types of networks are sometimes called gateways.

Figure 6.5 shows a picture of a router that is used in an DSL network. This router provides multiple Internet addresses for each customer that attaches their equipment to the DSL network..

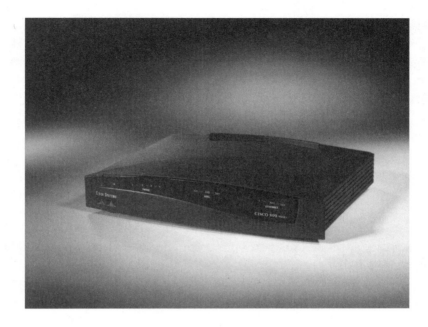

Figure 6.5, Cisco 827 Business ADSL Router

Source: Cisco

ATM Switch

An ATM switch is part of a packet data and switching system that rapidly transfers information by using fixed length 53 byte cells and high-speed communication links. ATM system use high-speed transmission lines (155 Mbps).

The ATM switching system is a connection-based system. When an ATM circuit is first established, a patch through multiple switches is setup and remains in place until the connection is completed. ATM service was developed to allow one communication medium (high-speed packet data) to provide for voice, data and video service.

As of the 1990's, ATM has become a standard for high-speed digital backbone networks. ATM networks are widely used by large telecommunications service providers to interconnect their network parts (e.g. DSLAMs and Routers). ATM aggregators operate networks that consolidate data traffic from multiple feeders (such as DSL lines and ISP links) to transport different types of media (voice, data and video).

The ATM switch rapidly routes packets to the pre-designated destinations. An ATM switch maintains a database (called a routing table). The routing table is updated each time a connection is setup and disconnected. This allows the ATM switch to forward packets to the next ATM switch or destination point without spending much processing time. The ATM switch also may prioritize or discard packets based on network availability (congestion). The ATM switch determines the prioritization and discard options by the type of channels and packets within the channels that are being switched by the ATM switch.

Network Management

A DSL service provider must be able to configure and operate their network equipment. These options include how they will manage the network equipment. An effective network management protocol is simple network management protocol (SNMP). SNMP is an industry standard communication protocol that is used to manage multiple types of network equipment. By conforming to this protocol, equipment assemblies that are produced by different manufacturers can be managed by a single program. SNMP protocol can operate via Internet protocol. The communication principles of SNMP are similar to those provide by IBM's NetView.

Chapter 7

xDSL Economics

The number of wired telephone lines worldwide in 2000 was over 700 million and this is projected to grow to 1.2 billion by 2002 [1]. The number of xDSL lines in service in North America totaled over 1 million at mid-year 2000 [2] and over 1 million by mid-year 2000. Of the total number of xDSL lines, the incumbent local exchange companies (ILECs) in the United States had 77 percent of the market share and competitive local exchange carriers (CLECs) had 22 percent market share. Long distance inter-exchange carriers (IXCs) accounted for less than one percent of the market share. At the end of 1999, Canada had 95,300 xDSL lines in service. A leading xDSL consulting firm TeleChoice estimates that the number of xDSL lines will grow to 2.1 million by the end of 2000 and will aggressively grow to over 9.6 million xDSL lines by the end of 2003 [3]. Figure 7.1 shows the number of installed DSL modems and projected growth of xDSL customers in the United States.

In 1998, the marketplace for fixed telecommunications services in the United States was estimated at $110 billion, with local telephony and data services accounting for approximately $47 billion [4]. Multimedia data services are the fastest growing portion of the telecommunications network. In addition to incumbent local exchange carriers (ILECs), other companies are starting to com-

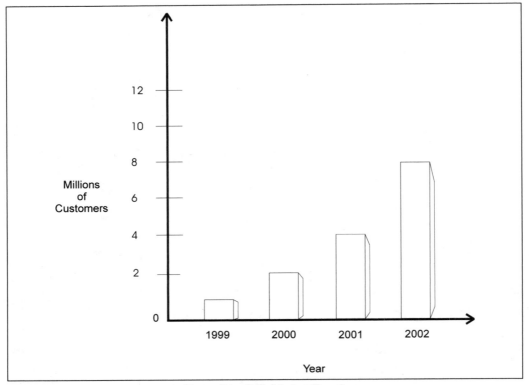

Figure 7.1, xDSL Market Growth

Source: Telechoice xDSL.com

pete for voice and high-speed data communications marketplace. These include cable television companies (CATV), wireless carriers, electric utility companies and new competitive access providers (CAPs).

Because of significant competition to high-speed communication lines, xDSL service providers are introducing advanced value-added services in addition to the high-speed data transmission features of xDSL technology. This allows the telephone service to create a competitive advantage while maintaining relatively high profit margins.

xDSL wholesale equipment costs have dropped over 15% per year between 1997 and 2000. While the technology and mass production cost reductions for telephones and systems are mature, new xDSL equipment is more complex than analog equivalents. However, when telephone equipment is produced in mass quantities, large sales volume promotes cost savings. Equipment costs for xDSL equipment must compete against a mature, competitive analog modem, which already has the advantage of cost reductions due to large production runs. However, xDSL systems offer significant advantages compared to analog systems and the economics of the network, from an access and service provider perspective, are very appealing.

The economic goal of a telephone and data network system is to effectively serve many customers at the lowest possible cost. The ability to serve customers is determined by the capability and capacity of the telephone system or systems. The key factors that determine the capacity of the system include the type of usage (holding time), sharing access methods, the efficiency of the transmission channels, and the number of lines installed for each node.

For any access technology, system capacity can be measured by the availability of service to customers when they desire to use it. The capacity of xDSL systems is determined by the capacity of the local loop channels (which xDSL line and technology conditions), capacity of network interconnection (capacity and protocol of interconnection lines), and the capacity of the systems the customers are connecting to (e.g. Internet or Frame Relay system).

The data transmission capacity of all three of these access systems is important to total system capacity. For example, if the xDSL service provider provides interconnection to the Internet using a 1.5 Mbps T1 line to provide service to 10,000 customers that each has ADSL modems, the capacity of the system will be severely restricted. This is called oversubscription. If the access provider interconnects these same customers through DSLAMs that share a single 10 Base-T interconnection system, the capacity of the inter-access network will be restricted. If the service provider does not remove bridge taps or loading coils from local lines that are used for ADSL technology, the local loop capacity will be limited.

Service providers strive to balance the system capacity with the needs of the customers. Installing a limited amount of data concentration equipment (DSLAMs) or low speed interconnection transport lines (limited backbone) results in unavailable or limited bandwidth communication channels for their customer. Installing systems that have excess capacity results in the purchase of system equipment that is not required. This increases cost to the end customer.

One of the key objectives of the new xDSL technologies is to achieve cost-effective service capacity, using techniques such as switch bypass, statistical multiplexing and more efficient use of copper lines and voice switching systems. A key advantage of offering xDSL service is to reduce the burden of the voice switching system through the redirection of data traffic via xDSL networks. In essence, an investment in xDSL equipment can reduce the required investment in voice digital switching system network equipment. The data calls that use voice switches for extended periods are re-routed through dedicated data networks.

Purchasing and maintaining system equipment is only a small portion of the cost of a telecommunication system. Administration, leased facilities, and tariffs may play significant roles in the success of voice and data communications systems.

In the late 1990's, the telecommunications marketplace began radical changes relative to new competition and discounted high bandwidth services. New service providers, such as cable television companies are competing in the marketplace with new competitive features, such as voice telephone services over coaxial cable lines. Wireless companies such as cellular and PCS companies are providing flat rate services that compete with landline services. Even power utility companies are beginning to offer telecommunications transport services through powerline carrier technology (discussed in chapter 4). All of these competitors are well positioned to provided local and long-distance telephone services. xDSL technologies are not needed for added services, they are needed for survival.

New advanced local loop technologies (such as xDSL) offer a variety of new features that may increase the total potential market and help service providers to compete. These new features may offer added revenue and provide a way to convert customers to a more efficient digital service. The same copper lines that provide voice services can offer advanced high revenue data transmission services. One of the key benefits of xDSL technology is for local telephone companies to gradually invest in xDSL technologies. This increases the barriers of entry for companies to offer high bandwidth service and has resulted in a very rapid market acceptance of xDSL technology.

End User Equipment Cost

The first xDSL commercial equipment was HDSL. HDSL modems were developed to replace T1 modems. HDSL modems offer the advantage of extended range, the elimination of costly line leveling (equalizing) and increased transmission distance without the use of repeaters (12,000 ft for HDSL compared to less than 6,000 feet for T1). The initial HDSL modems introduced in the 1980s cost over $5,000. In 1999, over 40% of the T1 lines used in the United States were transmitted via HDSL modems [5].

In the mid 1990's, ADSL modems were introduced. The average wholesale cost of ADSL modems has dropped from $750 in 1997 to approximately $500 in 1999 [6]. The production cost of ADSL modems is related to: development cost, production cost, patent royalty cost, marketing, post-sales support, and manufacturer profit. The cost of producing xDSL modems is decreasing due to economy of scale (volume production) and component integration.

Development Costs

Product development costs are non-recurring costs that are required to research, design, test, and produce a new product. Unlike well-established analog modem technology, non-recurring engineering (NRE) development costs for xDSL modems can be high due to the added complexity of high-speed data transmission design. Several companies have spent millions of dollars developing xDSL products. Figure 7.2 shows how the non-recurring development cost per unit varies as the quantity of production varies from 20,000 to 100,000 units. Even small development costs become a significant challenge if the volume of production of the xDSL modems is low (below 20,000 units). At this small production volume, NRE costs will be a high percentage of the wholesale price.

The introduction of a new technology within an industry presents many risks in terms of development costs. Some development costs that need to be considered include: market research; technical trials and evaluations; industrial, electrical, and software design; prototyping; product and government type approval testing; creation of packaging, brochures, user and service manuals; marketing promotion; sales and customer service training; industry standards participation; unique test equipment development; plastics tooling; special production equipment fabrication; overall project coordination.

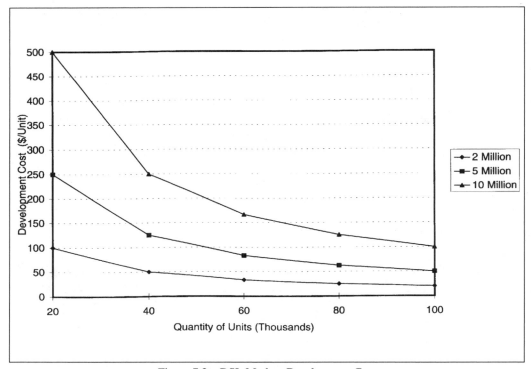

Figure 7.2, xDSL Modem Development Cost

Source: APDG Research

When a new product is rapidly developed for timely market introduction, it sometimes compromises a cost-effective design. Optimal cost-effective design is usually achieved through the integration of multiple electronic assemblies into a custom integrated circuit (IC). Custom IC chips can provide many circuits in one low-cost part. The development of custom Application Specific Integrated Circuit (ASIC) ICs usually require development setup costs that ranges from $250,000 to over $1,000,000. There may be more than one ASIC used in a xDSL modem.

Cost of Production

The costs to produce xDSL modems include; the component parts, assembly costs (factory production equipment and usage time), and assembly labor. xDSL modems are more complex than traditional analog modems due to their high frequency (RF like) design. AN xDSL modem is composed of an analog (audio processing and filtering) and digital signal processing sections.

The primary hardware assemblies that affect the component cost for xDSL modems are Digital Signal Processors (DSPs) and modulator/demodulator assemblies. A single DSP, or several may be used; it may cost between $7 and $28 [7]. The modulator and demodulator assemblies used in xDSL modems require multiple or higher frequency signals than those used in analog modems. Other components that are included in the production of an xDSL modem includes printed circuit boards, custom integrated circuits, discrete electronic components, filters, connectors, a plastic case or metal bracket for PC card mounting, status indicators, internal operating software (called firmware) and possibly other software (if used with a personal computer). In the year 2000, the total bill of materials (parts) for an ADSL modem was approximately $145 [8].

The assembly of xDSL modems requires a factory with automated assembly equipment. The factory assembly equipment for each production line usually costs between 2-5 million dollars. Dependent on the number of parts to be inserted and capability of production equipment, a single production line can produce a maximum of 500-2,000 units per day (150,000-600,000 units per year).

Production lines require regular maintenance. Production lines are often shut down one day per week for routine maintenance and two weeks per year for major maintenance overhauls which leaves about 300 days per year for the manufacturing line to produce products. The cost to own and operate production equipment is approximately 25% of the equipment value per year. This is due to the interest or lease cost (10-15% per year) and the equipment depreciation cost (10-15% per year). This results in a production facility overhead of $500,000 to $1.25 million per year for each production line. Figure 7.3 shows how the cost per unit drops dramatically from approximately $10-25 per unit to $1-3 per unit as volume increases from 50,000 units per year to 400,000 units per year.

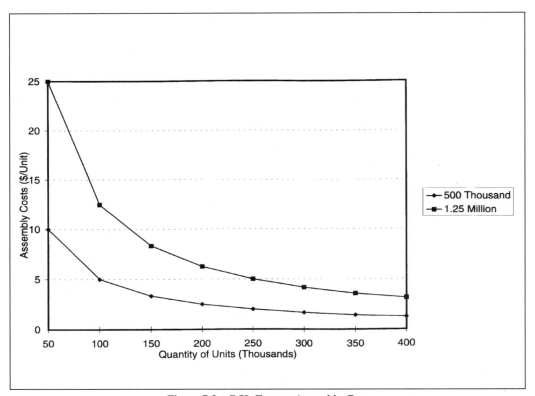

Figure 7.3, xDSL Factory Assembly Cost

Although automated assembly is used in factories for the production of xDSL modems, there are some processes that require human assembly, such as the placement of large bulky components (e.g., installation of large filters and metal brackets). Efficient assembly of a xDSL modem in a modern factory requires approximately 1/2 to one hour of human labor. The amount of human labor is a combination of all workers involved with the plant, including administrative workers and plant managers. The average loaded cost of labor (wages, vacation, insurance) varies from approximately $20-$40 per hour, based on the location of the factory and the average skill set of human labor. This results in a labor cost per unit that varies from $10-$40.

Patent Royalty Cost

Another significant cost factor to be considered is patent royalties. Several companies have disclosed that they believe they have some proprietary technology that is required to implement some features required or provided as options to the xDSL standard specifications. Many large manufacturing companies exchange the right to use their patented technology with other companies that have patented technology they want to use. Patents from other companies that may be essential to implement the standard specifications may be required. Some of these patents have been discovered or disclosed to the leading manufacturers.

Sales and Marketing Cost

Marketing costs are included in the wholesale cost of the xDSL modems and assemblies. These costs include the cost of direct sales staff, manufacturer's representatives, advertising, trade shows, and industry seminars.

xDSL equipment manufacturers often dedicate highly paid representatives or agents for key customers. Much like the sales of other consumer electronics products, manufacturers employ several technical sales people to answer a variety of technical questions prior to the sale.

Some xDSL manufacturers use independent distributors to sell their products. This practice is more prevalent for smaller, lesser-known manufacturers who cannot afford to maintain dedicated direct sales staff. These representatives commonly receive up to 4-8% of the sales volume for their services.

Advertising programs used by wireless radio manufacturers involve broad promotion for brand-recognition and advertisements targeted for specific products. The typical advertising budget for xDSL equipment manufacturers varies from approximately 3%-6%. The budget for brand recognition advertising normally ranges from less than 1% to over 4%. Product-specific advertising is often performed through cooperative advertising. Cooperative advertising is paid from the manufacturer to a distributor or retailer. The amount paid commonly varies from 2%-4% of the cost of the products sold to the distributor. For the distributor or retailer to receive the cooperative commission, they must meet the manufacturer's advertising requirements. This approach allows distributors and retailers to determine the best type of advertising for their specific markets.

xDSL equipment manufacturers ordinarily exhibit at trade shows three to four times per year. Trade show costs are high. The manufacturers that exhibit at trade shows normally have large trade show booths, gifts, and theme entertainment. Hospitality parties at trade shows are also common. These manufacturers often bring 5-20 sales and engineering experts to the trade shows to answer distributor questions. This is an expensive event as in addition to the staff payroll and travel expenses, the trade show costs include booth space rental ($10k to over $100k), booth display fabrication and shipping, and other costs.

To help promote the industry and gain publicity, manufacturers participate in a variety of industry seminars and associations. The manufacturers regularly have a few select employees who write for magazines and speak at industry seminars. All of these costs and others result in a combined estimated marketing cost for xDSL equipment manufacturers of 10%-15% of the wholesale selling price.

Post-Sales Support

The sale of xDSL equipment involves a variety of costs and services after the sale of the product (post sales support), including warranty servicing, customer service, and training. A customer service department is required for handling distributor and customer questions. Because the average end user of xDSL equipment is not technically trained in xDSL technology, the amount of non-technical questions can be significant. Distributors and retailers also require training for product feature operation and servicing. The post sales support cost for xDSL equipment is usually between 4%-6%.

Manufacturers' Profit

Manufacturers must make a profit as an incentive for manufacturing products. The amount of profit a manufacturer can make, as a rule, depends on the risk involved with the manufacturing of products. As a general rule, the higher the risk, the higher the profit margins.

The xDSL equipment market in the early 1990's became very competitive due to manufacturers' abilities to reduce cost through mass production and through competition from other high-speed modem technologies (e.g. cable modems). To effectively compete, manufacturers had to invest in factories and technology, which increased the risk associated with business success. In 1999, the estimated gross profit in the xDSL equipment manufacturing industry was approximately 10%-30% [9].

System Equipment Cost

The cost for xDSL system equipment includes the following primary factors: development cost, production cost, patent royalty cost, marketing, post sales support, and manufacturer profit.

Development Costs

The manufacturing of most xDSL systems equipment involves the development and use of new technology and the development cost of individual network system components (e.g. routers) can exceed millions of dollars. xDSL network system equipment development costs are much lower than local exchange network equipment development costs. This is because of the modular system approach that is based on established standards. However, due to system availability and reliability requirements, significant testing and validation is still required for new xDSL systems parts. This high investment can limit many manufacturers from producing a family of products for the xDSL systems.

While system equipment DSL modems ATU-C (central office) modems perform similar functions to subscriber ATU-R (remote) modems, the coordination of all the modems involves additional electronic subsystems. Additional assemblies include digital subscriber line access modules (DSLAMs), hubs or switching routers, and large databases to hold customer features and billing information. All these assemblies require hardware and complex software.

xDSL systems have a unique feature that has enticed many manufacturers to develop network products. For earlier high-speed telephone data transmission systems (e.g. ISDN), the equipment in the network was usually unique to the manufacturer and significant network changes were required. The network equipment manufacturer develops and changes components for an entire network, including the switching systems, software, DSUs, and controllers.

As an alternative, the specification xDSL transmission allows the use of many established network technologies (e.g. ATM and Ethernet). This means that many of the network parts in an xDSL system do not have to be unique to a single manufacturer. This allows a manufacturer to only develop parts of the network. Thus, the initial investment is lower and more manufacturers have produced system equipment. The increase in the number of network equipment component suppliers has decreased the cost of system equipment.

Similar to telephone networks, when a xDSL network system develops a problem, significant parts of the network or even the entire system can be affected. New xDSL hardware and software features require extensive testing. Testing xDSL systems can require thousands of hours of labor by highly skilled professionals.

Cost of Production

The physical hardware cost for xDSL network system equipment should be more expensive than the predecessor network system equipment (TSI based systems) due to the added technological complexity. However, the physical hardware cost for xDSL network system equipment may actually be less than older telecommunications equipment. This is a result of more standardized products and economies of scale.

The cost to manufacture a xDSL network system assembly (e.g. router, DSLAM) includes the component parts, automated factory equipment, and human labor. Because the number of xDSL modems that share a single DSLAM is typically 4 or more (a DSLAM is a concentrator) and several DSLAMs may be connected to a single network hub, the quantity of xDSL system assemblies produced is much smaller than the number of xDSL end user modems.

Setting up automated factory equipment is time consuming. For small production runs, much more human labor is used in the production of assemblies because setting up the automated assembly is not practical. The production of system equipment involves a factory with automated assembly equipment for specific assemblies. However, because the number of units produced for system equipment is normally much smaller than end user consumer products, production lines used for the manufacture of system equipment are often shared for the production of different assemblies, or the production line remain idle for periods of time.

With many carriers deploying xDSL systems, the demand for xDSL system equipment is increasing exponentially. This increased demand allows for larger production runs, which reduces the average cost per unit. Large production runs also permit investment in cost-effective designs, such as using Application Specific Integrated Circuits (ASICs) to replace several individual components.

The maturity of xDSL modem technology is promoting cost reductions through the use of cost-effective equipment design and low-cost commercially available electronic components. In the early 1990s, many technical system equipment changes were required due to changes in proprietary systems and specifications. Manufacturers had to modify their equipment based on field test results. For example, complex echo cancellers were required for some xDSL technologies.

In the early 1990s, it was also unclear which xDSL technologies would become commercially viable. This limited availability of standard components because integrated circuit manufacturers waited to see the marketplace developed before developing standard components. Today, the success of xDSL systems has created a market of low-cost digital signal processors and modems for xDSL systems.

Like the assembly of xDSL modems, the assembly of system xDSL modems and network routing equipment involves a factory with automated assembly equipment. The primary difference is there are smaller production runs for network equipment and each piece of network equipment may have multiple assemblies that are more complex then their end user equivalents.

The number of system equipment units that are produced is much smaller than the number of subscriber xDSL modems that are produced because each access node can serve several access lines (end customers). The result is much smaller production runs for xDSL network system equipment. While a single production line can produce a maximum of 500-2,000 assemblies per day, several different assemblies for xDSL access nodes (DSLAMs and routers) are required. A change in the production line from one assembly process to another can take several hours or several days. xDSL network equipment requires a variety of different connectors, bulky filter parts, and large equipment case assemblies. Due to the low-production volumes and many unique parts, it is not usually cost-effective to use automatic assembly equipment. For unique parts, there are a limited number of automatic assembly units available. Because of this more complex assembly, and the inability to automate many assembly steps, the amount of human labor is much higher than for end user xDSL consumer products.

The assembly equipment for each automated production line can cost two to five million dollars. The number of units that can be produced per day varies depending on: the speed of the automated component insertion machines, the number of components to be inserted, the number of different electronic assemblies per equipment, and the amount of time it takes to change/setup the production line for different assemblies. If it is assumed there are four electronic assemblies per xDSL access node equipment (e.g., controller, line modem, diagnostic processing section, and power supply), the automated production cost for network equipment should be over four times that of xDSL consumer products.

Figure 7.4 shows how the production cost per unit drops dramatically from approximately $400-$1,000 per unit to $50-$125 per unit as the volume of production increases from 5,000 units per year to 40,000 units per year. This chart assumes production cost is four times that of xDSL network equipment due to the added complexity and the use of multiple assemblies.

While automated assembly is used in factories for the production of xDSL network equipment, there are some processes that require human assembly. Efficient assembly of network units in a modern factory requires between five and ten hours of human labor. The amount of human labor includes all types of workers from administrative workers to plant managers. The average loaded cost of labor (wages, vacation, insurance) varies from approximately $20-40 per hour, which is based on the location of the factory and average workers skill set. The resultant labor cost per unit varies from $100-400.

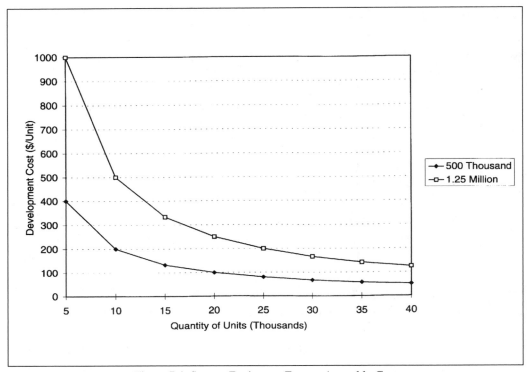

Figure 7.4, System Equipment Factory Assembly Cost

Patent Royalty Cost

Network equipment use many additional technologies than those used in xDSL consumer electronics (e.g. gateways and network switching). This usually involves using a portfolio of patents. Cross licensing among the large manufacturers is common and this tends to reduce the cost of patent rights. However, when patent licensing is required and no exchange is possible (some non-manufacturing companies thrive exclusively on patent portfolios), the patent costs are often based on the wholesale price of the assemblies in which the licensed technology is used. This increases the cost of equipment.

Marketing Costs

Marketing costs for system equipment include direct sales staff personnel, sales engineers, advertising budgets, trade shows, and industry seminars. xDSL system manufacturers often dedicate several highly paid representatives for key customers. Because system equipment sales are much more technical than the sale of xDSL consumer products, manufacturers usually employ skilled people to answer a variety of technical questions prior to the sale of network equipment.

Advertising used by the system equipment manufacturer primarily targets specific products. The budget for brand recognition advertising is commonly small and is targeted to specific communication channels. This is because the sale of xDSL system equipment is targeted to a small group of people who usually work for a telephone service provider. Budgets for product-specific advertising are targeted to industry specific trade journals. Much of the advertising promotion of brand name and system equipment occurs at industry trade shows. The total advertising budget for xDSL system equipment manufacturers is regularly less than 2%.

xDSL system manufacturers exhibit at trade shows normally three to four times per year. The system equipment manufacturers that exhibit at trade shows often have large hospitality parties that sometimes entertain hundreds of people. System equipment manufacturers typically bring 15-40 sales and engineering experts to the trade shows to answer key technical and business related customer questions.

To help promote the industry and gain publicity, system equipment manufacturers participate in many industry seminars and associations. These manufacturers use trained experts to present at industry seminars. All of these costs result in an estimated marketing cost for system equipment manufacturers of approximately 8-10% of the wholesale selling price.

Post-Sales Support

The sale of xDSL systems involves a variety of costs and services after the sale of the product. This includes warranty servicing, customer service, and training. A seven-day, 24-hour customer service department is required for handling customer questions. Customers require a significant amount of training for product operation and maintenance after a system is sold and installed. The post-sales support costs for wireless system equipment is ordinarily 3-5%.

Manufacturer Profit

Standardization of systems and components, particularly xDSL, has led to a rapid drop in the wholesale price of system equipment. While the increased product volume of system equipment has resulted in decreased manufacturing costs, the gross profit margin for system equipment has decreased. The estimated gross profit in the xDSL system equipment manufacturing industry is 10-15%.

Network Capital Cost

There are two basic options for network operators to offer high speed data networks; upgrade existing switching equipment to accommodate high speed data services or to create a separate data network and bypass the existing voice switching systems. In theory, existing public telephone switching technology could be upgraded to serve existing customers with high-speed data services. This is what ISDN systems were supposed to accomplish. However, expanding or upgrading existing switching systems requires significant investment to serve the small number of initial customers that have a high bandwidth need.

The telephone service provider's investment in network equipment includes xDSL modems, DSLAMS, a high-speed interconnection network, gateways (e.g. to the Internet), and network databases. One primary objective of the new xDSL technologies is to provide high-speed data access at a low network equipment cost per customer (capital cost per customer). xDSL makes this business objective possible as network equipment can be gradually added and customers can be billed based on the speed of their services.

One of the reasons that xDSL technologies were developed was to allow for cost-effective capacity expansion of the copper distribution system. Cost-effective capacity expansion results when a single copper pair can carry multiple types of information. This allows more customers to be served by a smaller number of copper lines.

The use of xDSL system technology also reduces the average voice switching cost per customer as data connections through the voice switching are dramatically reduced. The holding time of voice switching systems has increased due to long modem connections to the Internet. As a result, the capacity of switching systems had to be increased. The use of xDSL technology to off load the interconnection requirements on the voice switching systems reduces the average capital cost per customer for voice switching equipment.

Access Node/DSLAM

An access node (DSLAM) in a DSL system is a concentrator and adapter that converts the high-speed network data lines into one or more types of xDSL signals. An xDSL access node is composed of a high-speed communications line connection, DSU/CSU or ONU, an interconnection (switching) system, one or more DSLAMs, several xDSL modems (several may be installed into each DSLAM). The access node can occupy less than 80 square feet of space. It is preferable to locate an access node in or very near to the main distribution frame (MDF) in the central office (CO) facility. The environment for an access node must be secure (safe), have access to power supply and climate control, and may have other non-standard requirements.

To interconnect high-speed data lines to access nodes, it is necessary to install a DS1, DS3, ATM or other high-speed communications line and a line interface. Fortunately, access nodes are commonly located in the central office (CO) near high-speed trunks. This limits the distance necessary to install a high-speed line (often fiber). In addition, a Digital Service Unit (DSU)/Channel Service Unit (CSU) or Optical Network Unit (ONU) will need to be connected to the access node. The cost of a DSU or ONU ranges from $2,000 - $10,000.

The access node equipment and equipment frame must be installed. An average installation cost is approximately $5,000. In addition, a backup power supply (batteries) may be included ($2,000 estimated). Figure 7.5 shows the estimated cost for an xDSL access node without the backbone network switching equipment.

For a single network, several DSLAM access node equipments must be purchased and installed. Each DSLAM interface site usually has several xDSL line ports. To determine the total number of customers that can be serviced by an access node, the number xDSL line ports are multiplied by the number of customers that can be serviced per DSLAM. The number of DSLAMs that can be located at a specific access node is dependent on the speed of the communica-

Item	Cost x $1,000's
Equipment cabinet	$5
DSU or ONU	$5
Install High Speed Comm Line	$5
Equipment Installation	$5
Backup Power Supply	$2
Total	$22

Figure 7.5, Estimated Capital Cost per xDSL Access Node

tions line, type of customers that use xDSL service, and customers within the xDSL transmission distance (typically within 10,000 to 15,000 feet).

Because not all subscribers use every xDSL channel at the same time, xDSL systems that are used for data interconnection service regularly add 20-40 (30-average) customers per 1.5 - 10 Mbps of network interconnection (dependent on use). This is called the loading of the system.

If there are an average of 3 DSLAMs installed in a single CO site and each DSLAM contains 14 line cards that each supply service to 48 customers then the total number of customers that get xDSL service is 2,016. These 2,016 customers must share the capital cost for the provision, installation and running costs for the xDSL network.

System Operations Center

Access nodes sites are usually connected to a gateway through an intelligent switching system (called the "router"). An operations center must be located in a long-term location (10-20 years), usually near a Public Switched Telephone Network (PSTN) central office switch connection or gateway to the Internet. The building will contain network monitoring and gateway equipment. Commonly, customer databases are located in this operations center facility. The operations center commonly contains software that allows advanced services are available at additional cost.

DSL System Cost Case Study

DSL service providers are likely to deploy multiple technologies including HDSL, ADSL and SDSL. Because of space constraints in the central office, a DSL provider is likely to initially install a single rack of equipment in the central office. As systems develop, six or more racks of DSLAM equipment could be deployed at each central office. If more than one rack is deployed an additional piece of equipment, called a concentrator, will be needed to feed the data traffic to the ATM switch.

Figure 7.6 shows a table that details the cost of upgrading a central office to offer xDSL service. This table shows that with an investment of only $269,500, a DSL service provider can offer high-speed data and voice services to over 2,000 customers. This has an average capital cost per customer of less than $120 per customer.

The voice gateway shown here concentrates the voice frequency bands from the pairs and passes them to the PSTN. The prices shown above represent the minimum capital cost of providing services to 2,016 SDSL subscribers at a maximum rate of 2.3Mbps each.

Description	Max Capacity	Unit Price	Total Cost
DSLAM's	2,016 pairs (SDSL)	$15,000	$15,000
Power Supply		$7,000	$7,000
Test Module		$5,500	$5,500
Voice Gateway	48,000 pairs	$50,000	$50,000
ATM Backbone Switch	112 E1's	$180,000	$180,000
Installation		$12,000	$12,000
Total Cost			$269,500
Cost per Customer (2,000 customers)			$134.50

Figure 7.6, DSL Cost Case Study

Operations, Administration and Maintenance Cost

The costs of operating a xDSL system include leasing and maintaining communication lines, interconnection tariffs, billing, administration (staffing) and maintenance. The operational cost saving benefits of installing xDSL equipment includes a reduction of voice switching capacity, a reduction in the installation of new copper lines, reduction in the maintenance of existing copper lines.

Leasing and Maintaining Interconnection Lines

Similar to the sharing a copper line with multiple access channels, the number of customers that can share the cost of an interconnection communication line (loading of the line) varies with the type of service. For voice only customers who ordinarily use the phone for 2-3 minutes per call and make and receive 5-15 calls per day (30 minutes per day on average); approximately 120 customers can share a T1 (5 customers per voice path x 24 voice paths per communication line) or 150 customers per E1 (5 customers per voice path x 30 voice paths per communication line). The average office customers use the phone for approximately 60 minutes per day. For office usage, approximately 60 customers can be loaded onto a T1 or 75 for E1. In developed countries that have a significant percentage of residential customers that have Internet access, the average line usage is much higher 90 minutes to 2 hours per day. This decreases the number of customers that can share interconnection lines. For voice and analog modem access lines, approximately 48 customers can share a T1 line (2 customers per voice path x 24 paths per communications line).

High-speed Internet access requires a higher data transmission rate than voice customers. High-speed Internet access requires approximately 1-6 Mbps peak per customer. Because each customer is not likely to access the Internet at the same time and when they are accessing the Internet, the average data transmission is much lower (below 100 kbps), several high-speed Internet users can share the Interconnection line 10-40 for each T1 line. This results in approximately 24 customers sharing the bandwidth of each T1 line. However, if customers access the Internet simultaneously then the bandwidth has to be shared between them and this reduces the average data transmission rate.

Multicast applications such as near video on demand (NVOD) require higher interconnection paths than high-speed Internet access such as a DS3 or E3. A DS3 interconnection line carries 44.7 Mbps (672 DS0s) and an E4 line carries 34.3 Mbps (512 DS0s). Although each customer will use approximately 1.5 Mbps for each digital video channel, several customers will be watching the same digital video channel so the data transmission rate is shared (similar to one-way CATV systems). If each the average video-viewing period is 2 hours and 3-5 customers watch the same video channel, the average data transmission rate for customers will be approximately 250 kbps during the peak viewing periods. This results in each approximately 2-4 customers per T1 line or .2 customers per DS0.

Direct digital video access services such as video on demand (VOD) require a much higher interconnection path such as a DS3, E3, DS4, E4 or ATM line. A DS4 interconnection line carries 274 Mbps (4032 DS0s) and an E4 line carries 137.5Mbps (2048 DS0s). Each customer will use approximately 1.5 Mbps for each digital video channel. However, not every customer will access the network at the same time. If each the average video-viewing period is 2 hours, the average data transmission rate for customers will be approximately 500 kbps during the peak viewing periods. This results in each approximately 2-4 customers per T1 line or .1 customers per DS0.

The monthly leased line cost per customer is determined by dividing the monthly cost by the total number of customers. Table 7.7 shows the estimated monthly cost for interconnection charges of access nodes dependent on the type of service. The estimated monthly cost is based on 100% use of the communication lines. If the communication lines are not used fully (it is rare that communication lines are used at full capacity), the average cost per line increases. This table shows that the interconnection lines must have more than 50 times the capacity when offering digital video services compared to voice services.

Service Type	Type of Leased Line	T1 Line Cost per Month	Number of DS0 Channels	Customers per Chan	Customers Per Leased Line	Total Cost per Month
Residential Voice	DS1	$300	24	5	120	$2.50
Business Voice	DS1	$300	24	2.5	60	$5.00
Residential Voice + Analog Modem	DS1	$300	24	2	48	$6.25
High Speed Internet	DS1	$300	24	.5	24	$12.50
High Speed Internet	DS3	$2000	672	.5	336	$5.95
Near Video on Demand	DS3	$2000	672	.2	134.4	$14.88
Broadcast Video	DS3	$2000	672	.1	67.2	$29.76
Broadcast Video	DS4/ATM	$5000	4032	.1	403	$12.40

Table 7.7, Monthly Interconnection Line Cost

Interconnection Tariffs

xDSL systems are often connected to other networks such as the Internet. When xDSL systems are connected to other networks (through a gateway), there is often an interconnection cost that consists of a fixed access cost plus a usage cost.

The voice portion of the xDSL system must be connected to the PSTN. For the ILEC, the cost of switching equipment, voice interconnection lines and signaling system are distributed over the revenue from calls that originate and terminate in the LEC network. For competitive local exchange companies that handle voice services (as a reseller), these calls must be routed to and from the ILEC network. The ILEC receives revenue from the CLEC for calls that terminate in its network and the CLEC receives revenue from the ILEC for calls that terminate in its network (this is called reciprocity). Unfortunately, most CLECs only have a small portion of customers in the ILEC service area, which means that the CLEC usually pays the ILEC for termination of a majority of calls from their customers into the ILEC network. This cost is approximately ¾ to 1 cent per minute. The average residential customer originates approximately 30 minutes of calls each day, which results in a cost of approximately 25 cents per day or $7.50 per month.

In the United States and other countries that have separate long-distance service providers, when long distance service is provided through an ILEC, a tariff is paid from to the local exchange company (LEC). These tariffs can be up to 45% of the per-minute charges for long distance service. Due to government regulations limiting the bundling of local and long-distance service, it is necessary for some service providers to separate their local and long-distance service. Recent regulations may permit CLEC carriers to bypass the ILEC and save these tariffs.

Billing Services

Billing involves gathering and distributing billing information, organizing the information, and invoicing the customer. As customers initiate calls or use services, records are created. Each billing record contains details of each call that is to be billed. The details include who initiated the call, where the call was initiated, the time and length of the call, and how the call was terminated. Each call detail record (CDR) contains approximately 100-200 bytes of information [10]. The billing records are usually stored in the company's own database.

If the customer is allowed to access services of other companies (e.g. IXC or information services), billing records must be exchanged with these companies. Until the 1980's, telephone systems required the use of a magnetic tape to transfer records to standard Automatic Message Accounting (AMA) format. Today, billing records are regularly sent directly to each other or through a clearinghouse company to accumulate and balance charges between different service providers.

With the introduction of advanced services, billing issues continue to become more complicated. The service usage cost may vary between different information service providers. To overcome this difficulty, some information service providers (e.g. directory assistance) have agreed to bill customers at the billing rate established in their home telephone system.

Each month, billing records must be totaled and printed for customer invoicing, invoices mailed, and checks received and posted. The cost for billing services is approximately $4.50 per month. Billing cost includes routing and summarizing billing information, printing the bill, and the cost of mailing. To help offset the cost of billing, some telephone service providers have started to bundle advertising literature from other companies along with the invoice. To expedite the collection, some telephone service providers offer direct billing to bank accounts or charge cards.

Staffing and Maintenance of Access Lines

Running a telephone service company that provides voice and xDSL services requires staff workers with many different skill sets. Staffing requirements include executives, managers, engineers, sales, customer service, technicians, marketing, legal, finance, administrative, and other personnel to support vital business functions. The present staffing levels for telephone service providers is approximately 2.35 employees for each 1,000 customers [11]. If a loaded cost of $40,000 is assumed per employee (salary, expenses, benefits, and facility costs), this results in a cost of approximately $7.80 per month per customer ($40,000 x 2.35 employees /1,000 over 12 months).

The average number of employees is decreasing as maintenance and repair of systems becomes more automatic and digital systems are more reliable than early analog systems. Most new xDSL systems have automatic diagnostic capabilities to detect when a piece of equipment fails. These systems usually have automatic backup systems that can provide service until the defective assembly is replaced. Much of the cost reductions that have resulted from the decreased servicing cost have been reduced by the higher cost of advanced employee skill sets that are required to maintain more complex systems.

Service Revenue Potential

At the end of 1999, the average T1/HDSL access line revenue was $300 per month [12] and ADSL access line revenue in Q1 2000 for business customers was $125 (excluding back-haul revenues) [13] per month in the United States. The average revenue per residential subscriber in Q1 2000 was $68 (excluding back-haul revenues). The T1/HDSL access revenue US has declined approximately 15% each year over the last five years. This is primarily due to competitive pressure from other companies offering high speed data services such as cable companies, microwave carriers, utility companies with installed fiber lines. Outside the United States, monthly leased line revenue is higher but also decreasing.

The rapid growth of xDSL markets is also due to increased awareness of advanced xDSL services (e.g. Internet access and Frame Relay). Because many new xDSL systems are only starting operation, it is reasonable to assume a continued yearly growth of over 100% per year for most markets that offer xDSL service. The main revenue for xDSL service providers is derived from providing telecommunications service to businesses. However, in 1998, a significant portion of service revenue came from providing Internet access services to consumers. The average monthly revenue for xDSL Internet access is $68 per month.

Voice Service Cost to the Consumer

Since the mid-1980's, the average monthly revenue per access customer (analog line) in the United States has been approximately $20 per month. Some ILECs claim that the cost to provide voice service (analog lines) exceeds $40 per month. The difference in cost is paid for by high tariffs (approximately 45% of gross revenue) imposed on long distance service providers (IXCs) for having access to local customers through the ILEC system.

To help attract customers to xDSL service, some telephone companies are subsidizing the cost of xDSL equipment. Some xDSL service providers have also offered discounted Internet usage rates for customers that agree to high usage or long-term service plans.

Data Service Cost to the Consumer

There are two types of data services that are available to customers: continuous (called "circuit switched data") or brief packets (called "packet switched data"). Commonly, data transmission is charged at a fixed rate for a guaranteed data transfer rate and quality of service.

Internet Access

Customers purchase xDSL access based on their own value system, which estimates the benefits they will receive. A strong feature is the high access speed to the Internet. This single feature is often used for product differentiation and to increases information service revenue. In 1999, the average monthly analog line Internet access revenue in the United States was $15.50 [14] while the average monthly revenue for xDSL Internet access was $80 [15].

Other Services

A telephone service provider that has xDSL capability can offer access to other high value services. These services include high-speed data connection, multicast applications and digital video services. High-speed packet and data access services such as frame relay, switched multimegabit digital service (SMDS), ATM and other high-speed data access services. Multicast applications offer value added revenue potential. Multicast services include near video on demand (NVOD) and high-speed digital video conferencing service. Chapter 10 describes the applications that use these services.

Marketing Considerations

Marketing considerations for xDSL systems include high end user equipment cost, inability for all different xDSL equipment to inter-operate correctly, limited availability of xDSL equipment, system capacity limitations, high end-user disconnection rate (churn) and availability to distribution channels.

End User Equipment Cost

End user equipment cost is a strong consideration as significant entry into the consumer market occurs. The cost of DSL modems has been decreasing due to competition from other high-speed data services such as cable modems and microwave bypass. However, high end user equipment is a significant barrier for consumer market segment. Some xDSL carriers subsidize the price of xDSL modems or lease xDSL modems as a method to overcome the high initial barrier of modem equipment cost.

Interoperability Challenges

xDSL equipment that is manufactured by two different companies may not be completely compatible. Even if the equipment is manufactured to industry standards, industry standards are subject to interpretation and some features may be optional. Incompatibility can be in the form of lower data transmission performance or the inability for an application to operate (such as voice on the Internet). When the consumer experiences lower than expected data transfer rates or is not able to run specific types of applications, this may result in lower consumer confidence. Low consumer confidence leads to churn and bad publicity.

Availability of Equipment

The design and production of xDSL equipment requires significant investment by a manufacturer. xDSL modems are more complex than analog modems. Although xDSL systems define standardized products that have the potential for relatively large production quantities, the quantities of production for xDSL equipment is likely to be lower than analog modems for several years. This limits competition and is likely to reduce the availability of xDSL modems.

Capacity Limitations

Based on the economic considerations for high-speed interconnection line costs and the costs of high-speed gateways to other networks (such as the Internet), the carrier's xDSL backbone may not have the capacity to effectively serve all the new customers (over subscription). For example, is everybody in the network wants to access the Internet at the same time and connect to streaming web video (high data transmission) during a presidential address, the network is likely to be capacity limited and either some (or all) users will not get the access they desire.

Churn

Churn is the percentage of customers that discontinue xDSL service. Churn is usually expressed as a percentage of the existing customers that disconnect over a one-month period. Churn is often the result of natural migration (customers relocating) and switching to other service providers. Unlike voice services where there is a limited option for customers to switch, xDSL like services (high-speed data transmission) can be obtained from a variety of competing companies.

Because some xDSL service providers subsidize the equipment purchase to help reduce the initial cost of xDSL service, this can be a significant cost if the churn rate is high. Some xDSL carriers will go to various lengths to reduce churn including the signing of lengthy service agreements. These service agreements have a penalty fee in the event the customer disconnects service before the end of the period.

Distribution and Retail Channels

Products produced by manufacturers are distributed to consumers via several distribution and retail channels. The key types of distribution channels include: representatives, wholesalers, specialty retailers, retail stores, and direct sales.

Representatives (commonly called "reps" or "agents") are companies or people that sell xDSL or related products. Representatives may offer products from other companies and they are typically commissioned to sell products in a specific geographic region.

Wholesalers purchase large shipments from manufacturers and normally ship small quantities to retailers. Wholesalers will usually specialize in a particular product group, such as modems, answering machines, and electronic office products.

Specialty retailers are companies that focus on a particular product category such as electronics or communications. Specialty retailers know their products well and are able to educate the consumer on services and benefits. These retailers usually get an added premium via a higher sales price for this service.

Some xDSL service providers employ a direct sales staff to service large customers. These direct sales experts can offer specialty service pricing programs. The sales staff may be well trained and regularly sell at the customer's location.

Distribution channels are commonly involved in the service subscription process. The application for xDSL service can be a lengthy process as it may involve the determination of availability of xDSL service, a credit check for the customer (for advanced services) and the sale and activation of multiple xDSL modems.

Reseller Competitive Tactics

CLECs usually start with the competitive disadvantage of lack of facilities, staff and other assets that ILECs have. To gain access to a marketplace, CLECs often resell the services of ILECs. To help complete, CLECs often convert low cost lines to high-speed xDSL lines, offer high-speed Internet access and Internet telephony service.

Low Cost Leased Line Conversion

CLECs may lease a low cost copper line from an ILEC (called an alarm line) and install xDSL technology on each end of the line to convert the line to multi-channel capability. The new capacity offered by these upgraded lines can be sold as a premium service.

Internet Telephony

CLECs may promote low cost long distance telephone service through the Internet. If the DSL service provider includes a PSTN gateway and assists the customers with the installation of integrated access device (IAD) or Internet telephony software, the DSL provider may offer local telephone services without connecting to the LECs switching system.

Local Carrier Competitive Tactics

Although CLECs are customers of ILECs services, ILECs often view CLECs as fierce competitors. ILECs compete with CLECs by limiting the types of available lines and installing line conditioning on low cost lines to prohibit the use of xDSL equipment.

Lack of Line Availability

ILECs may limit the availability of low cost copper lines (alarm lines) in their network. CLECs often lease low cost copper lines and convert them to high capacity digital lines. These digital lines can compete with premium services offered by the LEC. Higher cost copper lines would increase the monthly copper cost to the CLEC. This would primarily affect low revenue lines offered by the CLEC (residential users).

Installation of Filters

The installation of line conditioning filters reduces the ability of competing telecom companies from purchasing audio copper lines at low tariff rates and converting them to xDSL technology. These loading coils attenuate the high frequency signals used by xDSL technology thus eliminating the ability of the line to be used as a high-speed data line.

References:

1. Globalstar 10K report, 31 March 1998, US Securities and Exchange Commission.

2. XDSL.com, August 15, 2000.

3. ibid.

4. 10K report, 31 March 1998, US Securities and Exchange Commission

5. Interview with xDSL industry expert, 15 Aug 2000.

6. ibid.

7. ibid.

8. ibid.

9. Interview with Wall Street analyst, 31 July 2000.

10. D.M. Balston, R.V. Macario, "Cellular Radio Systems," Artech House, 1993, pg. 223.

11. "1998 Annual Report," Ameritech, 1998.

12. T1/HDSL monthly revenue 1999.

13. Interview, Wall Street analyst, 31 July 2000.

14. ibid.

15. ibid.

Chapter 8
Future High Speed Digital Transmission Technologies

Improvements to the capacity of xDSL systems are only part of the future for xDSL systems. Advances include improved reliability of transmission, new protocol adaptations, additional bridges to alternative distribution mediums (e.g. power lines and home phone lines) and end-to-end digital telephone systems.

xDSL Transmission Improvements

xDSL technology evolved from the need to better utilize existing infrastructure and standard wiring systems. To improve the data transmission capacity of twisted pair, copper loop transmission systems can use advanced echo canceling, better line coding (modulation and data structure), increased bandwidth, improved cabling, and new transmission mediums (such as power line data transmission).

Advanced Echo Canceling

Echo signals on a communication line cause distortion that can reduce or completely inhibit the line to transmit data signals. Even in the existing DSL technologies, a reduction in the amount of echo distortion generally increases data transmission rates. Using advanced signal processing techniques and digital signal processor (DSP) integrated circuits, the ability to remove unwanted echoes has dramatically increased since the mid 1980's.

Echo exists in almost all transmission system. Echoed signals are delayed representations of the original signal that is transmitted. Echoes can be created by mismatches in electrical circuit impedance where some of the signal energy is reflected back to the signal source and line imbalances. The hybrid converters balance and remove the original signal from the audio portion of the received signal. If the hybrids are not balanced, echoes can result.

Some DSL systems are not as sensitive to signal echoes. These systems transmit and receive on different frequencies. However, to increase system capacity, DSL systems are likely to use simultaneous transmission through the hybrid circuits to take advantage of all the available bandwidth.

Two wire telephone systems carry information in two directions on only two wires. In some systems, the system uses two different frequency bands. The use of two different frequency bands reduces the impact of echo distortion. However, some xDSL systems transmit information in both directions in the same frequency band.

Figure 8.1 shows an advanced echo canceling system. This diagram shows that a signal is sent from a transmitter to a receiver that experiences distortion due to signal echoes. The echoes are created from an imbalance in the hybrid assembly and the echo is a delayed resemblance of the original signal that is delayed a few milliseconds. The amplitude of the delayed (echoed) signal is 10 dB below the original signal (10% of the original signal level). The delayed signal is combined with the original signal as it is transferred and this causes distortion. At the receiving end, an advanced canceling signal processor samples the received signal. This diagram shows that an intelligent echo canceling system continuously samples the signal for brief periods, delays the sampled signal, adjusts the sampled signal's amplitude, subtracts the suspected echo signal and determines if an echo signal is present. If the echo canceling system determines there is an echo, the amount of time delay and level of echo is stored and the subtraction process continues.

Until recently, the digital signal processing complexity of echo canceling limited the ability of systems to use echo canceling. Low cost, high performance digital signal processors (DSP) that use advanced echo canceling software from value added resellers (VARs) can more than double the data transmission rate by eliminating distortions caused by unbalanced lines for some DSL systems.

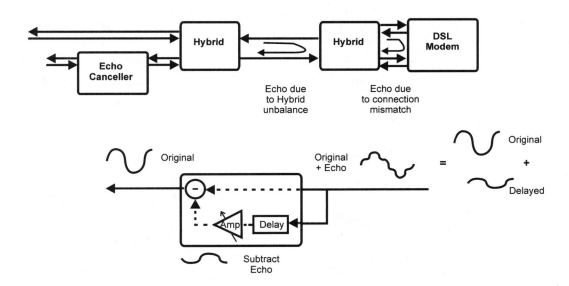

Figure 8.1, Advanced Echo Canceling

Improved Line Coding

Modulation is the process of changing the amplitude, frequency, or phase of a radio frequency carrier signal (a carrier) to change with the information signal (such as voice or data). Modulation efficiency is the amount of information that can be transferred onto a carrier signal within a limited bandwidth. The higher the modulation efficiency, the large the amount of information that can be transferred in a particular amount of bandwidth. To increase the modulation efficiency, DSL systems use a combination of phase and amplitude modulation.

Figure 8.2 shows an advanced modulation structure that allows a single symbol to represent several discrete levels. By adding additional decision points, more bits can be represented and a high data transfer rate is possible. This diagram shows that each decision point is a combination of phase and amplitude levels.

The difficulty with having more discrete levels is that even small distortion levels from other cables (signal ingress) or radio signals can cause errors. Distortion (such as amplitude spikes or phase delays) may indicate an incorrect decision point. Because each decision point represents several digital bits, this results in data transmission errors. As a result, there is a tradeoff between modulation efficiency (line coding) and ability to operate in poor transmission environments.

Increased Bandwidth

Increasing the transmission bandwidth increases the potential data transmission rate. It is possible to send frequencies that are higher than 1 MHz on twisted pair wires. High frequency transmission above 1 MHz is used in premises distribution networks (PDN) for short distances. These systems have been shown to

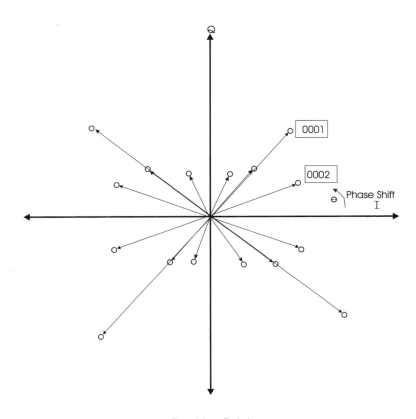

o- Decision Points
Figure 8.2, Improved Line Coding

transfer over 50 Mbps for short distances.

Unfortunately, the frequency response of twisted pair wire dramatically rolls off at high frequencies (above a few hundred kHz). High frequency attenuation of a twisted pair can be partially overcome by shorter distance cables and pre-emphasis and de-emphasis circuits. However, other challenges that occur when high frequency transmission (up to 20 MHz) are transmitted on a twisted pair. This includes interference to and from other electronics devices, substantial differences in transmission line characteristics due to installation differences. This will likely limit the use of high bandwidth DSL to very short distances in low RF interference areas (e.g. residential neighborhoods).

Improved Cabling

Using improved cabling can dramatically improve the characteristics transmission medium (e.g. reduced attenuation to high frequencies) and reduced emission and susceptibility to interference sources. Improved cabling can offer a wider frequency bandwidth and reduction in signal leakage. New types of cabling (such as coax cable) can be used for local telephone distribution.

Powerline DSL

Electric power lines are capable of carrying data signals up to 3 Mbps [1]. Electric utility companies have a tremendous advantage of "right of way" for cabling. Since the 1980's, many power companies have been replacing some of their grounding cables with new cables that have a fiber optic core. This provides tremendous data transmission bandwidth opportunities.

Due to regulatory issues, the fiber optic communication systems were initially used for internal company communications. This kept the electric utility companies from entering into or competing with the telecommunications industry.

As regulations eased with the general trend of deregulation (and increased competition), some of the electric utilities have installed fiber cables and some of the transmission capacity of these fiber links has been sold to long distance (intersystem) service providers. Recent deregulation of the telecommunications industries is starting to allow electric power companies to offer some communications services directly to customers.

Figure 8.3 shows a Powerline Data Distribution system. This diagram shows that data transmission paths are divided between high voltage (e.g. 10,000 volts+), medium voltage (less than 2,000 volts) and low voltage (250 volts or less). It is easier to install fiber optic cable on the high voltage lines than to share data in high voltage cable. Fiber optic cables are not as susceptible to the large amounts of electromagnetic interference that is produced by high voltage signals. Medium voltage systems (downstream of high voltage transformers) produce less interference, however the power lines act as antennas that can transmit and receive high frequency data signals.

One of the most appealing parts of powerline data transmission is the local drop and distribution through power lines in the home. It has been demonstrated that data transmission from the local transformer to the home can achieve over 3 Mbps. If an access node is installed downstream of the transformer (on the side closest to the homes), the 3 Mbps can be received by all the homes attached to the transformer. Each home that wants to access the data installed a gateway (data tap) near their electrical distribution panel. This gateway provides a standard data channel (such as Ethernet).

Although 3 Mbps does not seem like such as high data transmission rate if it is shared by multiple users, in some undeveloped countries power lines are the only installed cables to residential areas. The use of power lines to distribute data would allow power customers to gain access to the Internet and have digital telephone service.

Another recent development is the sending of high-speed data through the powerlines in a home or a small business. It has been demonstrated that over 20 Mbps through home power line distribution is possible. Home powerline distribution converts the disadvantage of rapid signal attenuation at high frequency and the blocking effect local transformers have on high frequency signals works to a data transmission benefit. High frequency signals can be applied to the power system through outlets. These high frequency signals travel throughout

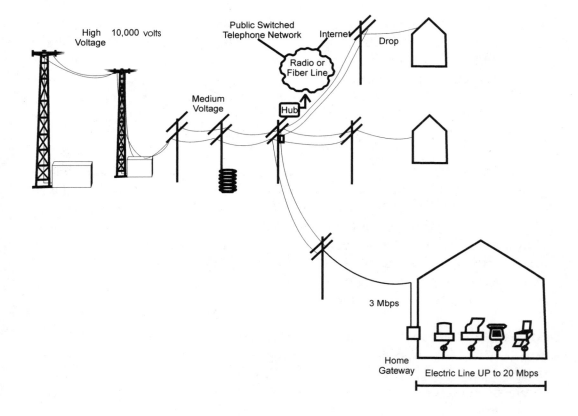

Figure 8.3, Powerline Data Transmission

the outlets in the house or business office. Other data devices receive the high frequency signals and convert them back into the original high-speed data signals. Because the high frequency signals do not travel far, they have little effect on nearby homes that are connected to the same power lines.

Figure 8.3 shows a powerline distribution system. This diagram shows that fiber cable is installed on the high voltage and medium voltage, power distribution grids. Because most of these lines are above ground and the towers are already constructed, installation of fiber cable is easy and low cost. As the fiber runs past a local (neighborhood) transformer, an access node is installed. This access node contains an optical network unit (ONU) that converts the fiber signal to a 3 Mbps signal that is pass onto the local drops to nearby homes. Optionally, a radio receiver could be used to supply the data signal for low voltage distribution. The 3 Mbps signal is received at a gateway in the home. This diagram shows a home powerline distribution network is part of the gateway. This converts the data signal into high-speed premises distribution network (PDN) that uses the home power lines. Optionally, the gateway could simply provide a data communications port such as Ethernet or include a home phoneline network gateway.

Consumer Electronics Bus (CEBus)

An existing data transmission system for home power lines is the consumer electronics bus (CEBus). CEBus is an industry standard for home networking that allows for data transfer rates up to 10 Mbps. Although CEBus technology is not suitable for long distance data transmission, CEBus technology does allow for broadband data transfer through the home wiring of a house.

Home Telephone Line Distribution

Similar to digital subscriber line technology that is used in the local loop, twisted wire pairs in the home can also be used to transfer high speed computer network signals.

Figure 8.4 shows a home networking system. This diagram shows that a high-speed data signal enters the home through a DSL bridge. The bridge converts the DSL modem signal to a high frequency signal that is applied to the home

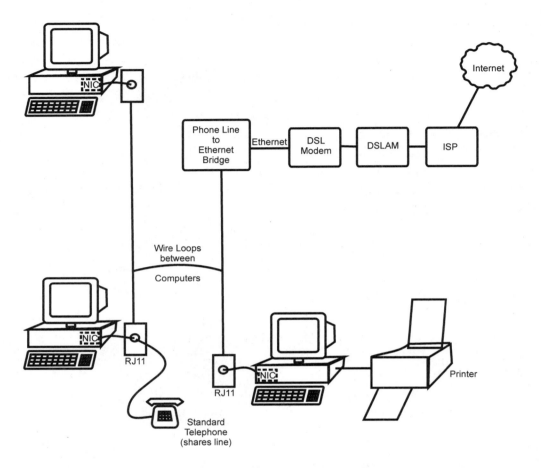

Figure 8.4, Home Networking through Telephone Lines

telephone wiring. Each data device that is connected to the phone jacks inside the home contains a phoneline network interface card (NIC) or data adapter. This converts the high frequency data signal into a data format that can be used by the computer (e.g. USB or PCI). The phoneline network uses a daisy chain format that allows multiple data devices (such as additional computers and printers) to be added without the need for a hub or router.

Reference:

1. Personal interview, ASCOM Corporation, industry expert, 10 April 2000.

Chapter 9

DSL Services

DSL services are standardized data communications protocols that connect customer premises equipment (CPE) to other communication devices (e.g. another CPE or a network database). Services allow customers to connect to systems such as the Internet, private networks or DSL services provide.

The communication network can be divided into several parts; local access lines (copper), voice communications network (PSTN), high-speed access network (xDSL), aggregator (interconnection), internet service provider (ISP) and content provider (media source). Figure 9.1 shows the different parts of a DSL network. Although this diagram shows several different service providers, a single service provider may be responsible for several parts of the system. For example, an existing telephone company (incumbent LEC) may be responsible for the copper access lines, high-speed backbone system aggregation and Internet service provider (ISP).

The different types of services enabled by high speed DSL include Voice, Video, and data. Bearer services (simple data transport) include virtual private networks (VPN), leased data line (e.g. T1 and E1), Point-to-Point Protocol (PPP) and ATM services can be provided by xDSL transmission technology.

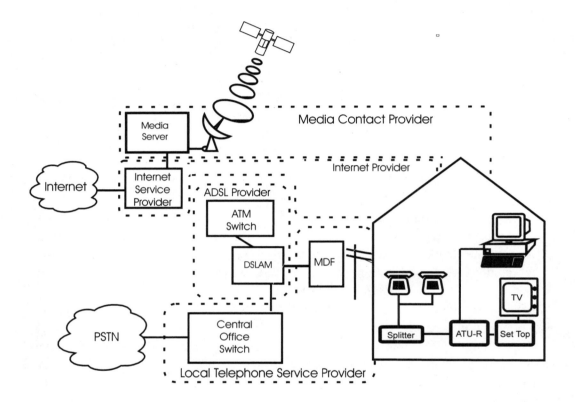

Figure 9.1, Telecommunications Network and Service Division

Voice over DSL (VoDSL)

There are two ways to combine voice over DSL service, frequency division and time division. Frequency division allows analog telephone signals to be sent over a DSL line on a dedicated low frequency portion (frequency band) of the line. The remaining higher frequency portion of the line (up to 1.1 MHz) is used to send digital information. ADSL, RADSL and VDSL allow frequency sharing of analog and digital signals.

Sending voice over a digital subscriber line system (VoDSL) is a process that sends audio band (also called "voice band") signals (e.g. voice, fax or voice band modem) via a digital channel on a digital subscriber line (xDSL) system. VoDSL requires conversion from analog signals to a digital format and involves the formatting of digital audio signals into frames and time slots so they can be combined onto a digital (DSL) channel.

To communicate voice signals via a DSL system, the customer requires one or more communication devices that are capable of converting audio signals into digital channels that can be transmitted via the DSL network. There must be another device that is capable of receiving the digital channel on the DSL system and converting the channel back into its analog voice band signal. These devices can be as simple as a computer with a sound card, a DSL modem and VoDSL software or as complex as a company's telephone network that has an integrated access device (IAD). Optionally, some DSL systems have a PSTN gateway that can convert digital audio on a DSL system into telephone signals that can be sent through the public switched telephone network.

VoDSL service allows a single digital system to provide both voice and data communications. VoDSL service bypasses the traditional public switched telephone network (PSTN). Because the voice signal is already in digital format, it can be routed through a network in a similar method as data. However, digital voice signals usually are coded with a different Quality of Service (QoS) requirement. Although digital data can often be delayed and retransmitted, digital voice signals must be transmitted in near real time (less than 100 msec of delay) to achieve acceptable quality levels.

Figure 9.2 shows a DSL network that is capable of offering voice services. This diagram shows that a single digital DSL line (such as HDSL, SDSL or ADSL) connects a customer to a DSL service provider. There are several standard voice band telephone devices shown in this diagram; an analog telephone and a fax machine. All three of these devices are connected to an integrated access device (IAD). The IAD converts the analog signals from the standard telephones and digital signal from the computer telephone into a format that can be transferred

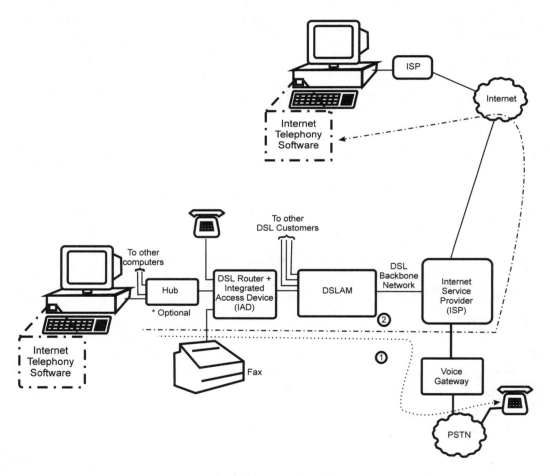

Figure 9.2, VoDSL Service

on the digital portion of the DSL line. The digital channels created by each VoDSL call may be routed to different places. If the call is a voice over Internet (VoIP), it could be routed to the ISP and the DSL system has completed its task. If the call is originating and terminating in the DSL network, the DSL system may route the digital audio to another DSL user. Optionally, the digital audio may be routed through a PSTN gateway so the call can be terminated to any other telephone that is connected to the PSTN network.

There are three basic types of configurations for voice over DSL; computer-to-computer, computer-to-telephone and telephone-to-telephone.

Computer-to-Computer

It is possible to originate and receive telephone calls between most personal computers that are connected to a DSL network. A computer-to-computer VoDSL call requires two (or more computers) that have a sound card, microphone and speakers. Because many personal computers come with a sound card, microphone and speakers, the installation may be simple and the only additional software necessary can be the software installation. The software essentially formats and routes the digital audio from the sound card to a data connection (such as the Internet). The data channel continuously routes the data to another computer that has similar computer-to-computer telephone processing capability.

Computer-to-Telephone

It is possible to originate and receive calls from a computer that is connected to a DSL network if the DSL network has a PSTN gateway and the customer has the appropriate DSL telephone equipment. A PSTN gateway is a communications device or that transforms data that is received from one network (such as the Internet or DSL network) into a format that can be used by the PSTN network. The PSTN gateway may be a simple device that performs simple call-completion and adaptation of digital audio into compatible signals for the PSTN.

Or the gateway may be a more complex device that is capable of advanced services such as conference calling, call waiting, call-forwarding and other PSTN like services. The PSTN gateway must create signaling protocols and compensate for timing differences between end users computer and the public switched telephone network (PSTN).

Telephone-to-Telephone

Telephone to telephone service over DSL requires interfaces at both the end users DSL termination and a PSTN gateway in the DSL system. For telephone-to-telephone service, an integrated access device (IAD) may be used. The IAD provides the necessary conversion from a customer's standard telephone equipment (e.g. POTS or ISDN telephone) to a digital channel on the DSL line. The DSL provider then routes calls via a PSTN gateway. This allows standard analog telephones can originate and receive calls without connecting through a local exchange company's (LEC's) voice switching system. The digital voice channels are then routed through the DSL network until they reach the PSTN gateway.

Video over DSL

Video over DSL is used for video conferencing, video on demand (entertainment) and other multicast and broadcast video services. For video conferencing, the customer should have at least 384 kbps to obtain reasonable quality video images [1]. For video on demand (such as digital movies), the customer should have at least 1.5 Mbps.

Because most consumer types of DSL (e.g. ADSL and VDSL) offer data transmission speeds that are different for the downstream and upstream channels, the end user must ensure that the data transmission rate is high enough in both directions if they use two-way video services (e.g. video conferencing).

For video conference calls, the IP address is used to receive calls. Basically, originating computer uses the IP address like a telephone number. The originating videoconferencing computer sends a message (calls) to the destination computer. The destination computer rings like a telephone. If the destination computer user is available and accepts the call, the video call is may start. Both computers then display the video images of both parties.

When a customer uses digital video for one-way viewing (e.g. video on demand), a high-speed data channel is only needed from the network to the customer. These channels may be selected from a nearby server that acts as a multicast hub.

Digital video normally requires a data transmission rate in excess of 50 Mbps. To enable digital video to be sent via xDSL systems, the digital video is encoded (compressed). The most common digital video compression system is the motion picture experts group (MPEG) version 2 digital video encoding. MPEG is an international standards organization (ISO) working committee that defines standards for the motion picture digital video compression and decompression for use in computer and digital broadcast systems. There are various levels of MPEG compression; MPEG-1 and MPEG-2. MPEG-1 compresses by approximately 52 to 1. MPEG-2 compresses up to 200 to 1. MPEG-2 typically provides digital video quality that is similar to VHS tapes with a data rate of approximately 1 Mbps. MPEG-2 compression can be used for HDTV channels, however this requires higher data rates.

Sound must also be sent along with the video. Fortunately, sound does not require as much data bandwidth as video and there are various audio compression techniques that dramatically reduce the audio bandwidth requirement. However, the data transmission of the audio will reduce the amount of bandwidth available for video. During a video conference call, it is possible to use another telephone line (e.g. the audio portion of the DSL line) to remove the burden of the audio processing and audio data transmission from the video conference call. This was common practice for low speed data connections such as 56 kbps

Internet connections. However, DSL systems have a much higher data transmission capacity so the separation of digital audio from digital video is usually not necessary.

There are various video conferencing standards including the International Telecommunications Union (ITU) H.323 and standard T.120 for multipoint data conferencing.

Video conferencing standards allow for the use of whiteboards. Whiteboards are devices that can capture images or hand drawn text so they can be transferred to a video conferencing system. Whiteboards allow video conferencing users to place share documents, images and/or hand written diagrams with one (or more) video conference call attendees.

Figure 9.3 shows the basic operation of sending video over a DSL connection. This diagram shows a computer with video conferencing capability that calls a destination computer. Computer #1 initiates a video conference call to computer #2 using the address 123.45.678.90. When computer #2 receives a data message from computer #1, a message is displayed on the monitor and an audio tone (ring alert) occurs. If the user on computer #2 wants to receive the call, they select the answer option (via the mouse or keyboard) that is generated by the software. Computer #1 then initiates a data connection with computer #2. The video conferencing software and data processing software in the computers (e.g. USB data bus and sound card) convert the analog audio signal from the microphone and digital video signal into a digital form that can be transmitted via the data link between the computers.

Most consumer video conferencing cameras use the universal serial bus (USB) connection to allow quick setup of video conferencing. Digital video transmission normally requires a fixed Internet address to allow computers to receive incoming video conference calls. As a result, video conferencing systems can have difficulties with intelligent networking devices (such as routers or firewalls and network address translation (NAT) devices). As discussed in previous chapters, some DSL Internet service providers use dynamic assignment of an IP address. This means that the address of the destination computer continually

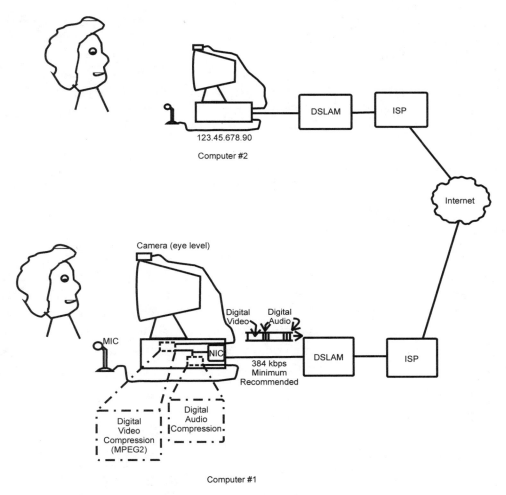

123.45.678.90

Computer #2

Figure 9.3, Video Services over DSL

changes and this can be a problem for video conferencing. One of the ways to solve the issue of dynamic IP addressing for video conferencing is for each of the users to log onto a video conferencing server that allows other people to reach you.

It is possible to overcome the challenges of dynamic assignment of an IP address by using a video conferencing web server. When two (or more) video conferencing workstations are connected to a video conferencing web server, the web server acts as the bridge between the users. To initiate a video conference call, one of the users establishes an account on the video conference call web server. When the first user logs onto the site, they provide a web address where the participant or participants can connect their computers.

Other video services over DSL use various accessories and software to capture and distribute digital video. These include WebCams and video mail software.

WebCam

Web Cameras ("WebCams") are digital cameras that provide images to the Internet. Some WebCams provide continuous image updates while others refresh their images every few seconds to reduce the data transmission requirement. WebCams can be inexpensive cameras that are mounted at eye level on top of a computer monitor or high quality rugged camera that are weather proof for mounting on buildings or poles.

Video Mail (VMail)

It is possible to send video mail (short video clips) via DSL links. Sending and receiving video mail involves the recording and sending short video clips. These video clips are typically 1-2 minutes long. Video clips may be sent via standard electronic mail (email). Video mail messages may be sent as an attachment to standard Email addresses.

T1/E1 Service over DSL

T1/E1 are standard digital transmission lines that are divided into twenty-four or thirty 64 kbps channels (commonly called voice channels). A T1 line (also called a DS1 line) is divided into a 193 bit frames and transmitted at 1.544 Mbps. DS1 signals can be transmitted in an unframed form where frames are 192 bits at a data transmission rate of 1.536 Mbps. The original use for DSL service (HDSL) was to support multi-channel (multiple DS0 channels) lines to emulate a standard T1 line.

A single pair of copper wires allows multiple 64 kbps digital DS0 voice channels (24 for T1 and 30 for E1). This diagram shows that each 64 kbps channel is combined in time sequence by a multiplexer into a standard high-speed digital signal (T1 or E1 format). The T1 or E1 signal is supplied to a HDSL2 modem. The HDSL2 modem converts the high-speed digital signal into a complex analog modulated signal that is sent onto the copper wire pair. On the receiving end, a HDSL2 modem converts the complex audio signal back into its high-speed digital T1 or E1 signal.

Virtual Private Networks (VPN) over DSL

Virtual private networks (VPNs) are secure private communication path(s) that connect one or more data networks via dedicated communication links between two points. VPN connections allow data to safely and privately pass over public networks (such as the Internet). The data traveling between two pints is encrypted for privacy.

There are several different types of VPN protocols. Popular VPN protocols include Point-to-point tunneling protocol (PPTP), layer 2 tunneling protocol, and Internet protocol security (IPSec). PPTP software is included with Microsoft Windows 95, 98, NT, and 2000 and is built into some routers. L2TP is an evolution of point-to-point tunneling protocol that offers more reliable operation and enhanced security. L2TP is part of Windows 2000. IPSec is part of the Internet

Protocol that helps to ensure the privacy of user data. IPSec is part of the next generation Internet, IPv6.

Figure 9.4 shows a virtual private network that is connected using DSL and the Internet. In this diagram, a remote computer is connected to a company communications network (hub) through a virtual private network (VPN). The computer first converts the data port (through a NIC) to a DSL modem. The data from the DSL modem is routed to the Internet. The Internet terminates the con-

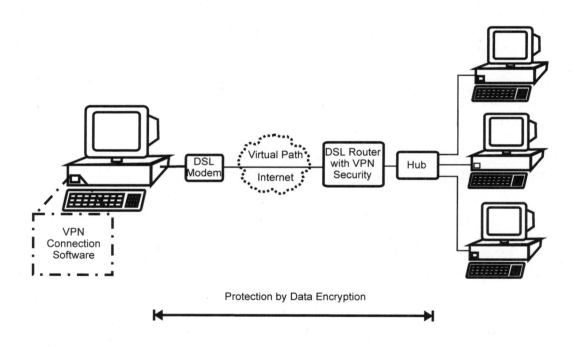

Figure 9.4, Virtual Private Network Service (VPN) over DSL

nection into the company's data network. Software that is installed in the remote computer and the corporate network adds data encryption to the data that is transferred between the remote computer and the corporate network.

There are a variety of after market products that offer enhanced security to VPN. These can be simple software only solutions to elaborate firewalls that analyze and predict unauthorized access. Various security options are described in chapter 5.

Point-to-Point Protocol (PPP) over DSL

Point-to-point (PPP) protocol uses a combination of Transmission Control Protocol (TCP) and Internet Protocol (IP) to allow end users (end points) to connect directly to the Internet via a communications connection (usually dialup connection). For Microsoft operating systems, this type of connection is managed by Dial-up networking (DUN). DUN software instructs the computer on how to initiate, maintain and disconnect data communications sessions with the Internet. Although DSL provides a continuous connection, some DSL service providers use PPP over DSL to provide connections. This means that the DSL service provider may require that Microsoft Windows be setup with DUN.

ATM over DSL

ATM is a packet data and switching technique that transfers information by using fixed length 53 byte cells. The ATM system uses high-speed transmission (155 Mbps) and ATM uses a connection based routing system. Connection based means that a same communications path through packet switches is established and maintained throughout the communications session. ATM service was developed to allow one communication medium (high speed packet data) to provide for voice, data and video service.

As of the 1990's, ATM has become a standard for high-speed digital backbone networks. ATM networks are widely used by large telecommunications service providers to interconnect their network parts (e.g. DSLAMs and Routers). ATM aggregators operate networks that consolidate data traffic from multiple feeders (such as DSL lines and ISP links) to transport different types of media (voice, data and video).

Initially, ATM service via DSL only allowed one virtual path connection (VPC) for an end user. Because the end customer for xDSL service is likely to have multiple data transmission requirements (e.g. digital video, high speed Internet and voice over DSL), ATM over DSL evolved to allow multiple virtual channels. ATM service allows multiple types of connections. These include constant bit rate (CBR), variable bit rate (VBR) and unspecified bit rate (UBR) services.

Constant bit rate (CBR) is a class of telecommunications service that provides an end user with constant bit data transfer rate. CBR service is often used when real time data transfer rate is required such as for voice service. Variable bit rate (VBR) service provides a variable data transmission rate of service to end user applications. Applications that use VBR services usually require some real-time interactivity with bursts of data transmission. An example of a VBR application is videoconferencing. Unspecified bit rate (UBR) is a category of telecommunications service that provides an unspecified data transmission rate of service to end user applications. Applications that use UBR services do not require real-time interactivity nor do they require a minimum data transfer rate. An example of a UBR application is Internet web browsing.

Because ATM uses a single fixed size of packets and has its own protocols, software interfaces between ATM service and applications must be created. These are called ATM adaptation layers (AAL). AALs convert the applications data messages into the very small packets that are transferred ATM and creates the necessary logical data channels.

In the early stages of ATM technology development, ATM was primarily used in commercial applications. The high-speed data communications of 155 Mbps was found to have more complex implementations than anticipated. To assist in the commercial rollout of ATM into computer networks, a low speed 25 Mbps version of ATM was created. Due to the popularity of the Ethernet system, the ATM25 standard became unpopular and limited amounts of commercial equipment were produced. However, the introduction of DSL into the home and the need for combined high-speed multimedia channels that offer varying levels of quality of service (QoS) has again increased the popularity of ATM25.

Figure 9.5 shows ATM service over DSL. This diagram shows that ATM can provide several different types of services. AAL5-rt provides a real time ATM

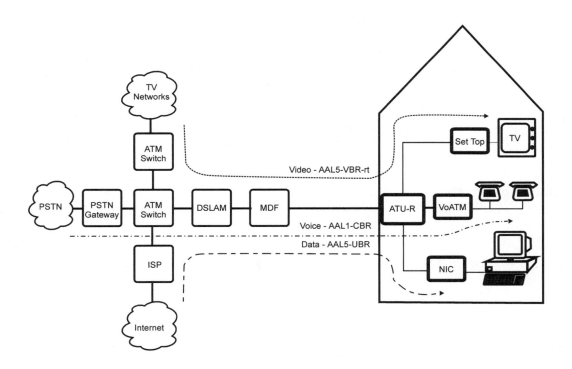

Figure 9.5, ATM over DSL

adaptation layer that is commonly used for digital video services that vary in bandwidth and require near real time connectivity. AAL1-CBR is a constant bit rate channel that is used for digital voice communications. AAL5-UBR is an unspecified bit rate channel that provides data on an as-available basis. The AAL5-UBR is an effective channel that allows for web browsing.

Reference:

1. David Angell, "DSL for Dummies," IDG, 2000, pg. 273.

Chapter 10

Broadband Applications

Broadband applications are software programs that require communication technology that offers data transfer speeds that can exceed 1 Mbps. Many of the communications applications and services that were available in the year 2000 were designed for the narrow-band applications (below 56 kbps). These applications included limited graphic web browsing, text based on-line shopping, email and word processor file transfer. Low cost broadband services such as xDSL systems provide a tremendous opportunity for the development of richer, more enhanced applications. These applications require services such as streaming video, rapid image file transfer or high speed data file transfer services.

There are hundreds of applications that are driving the demand for high-speed data transfer services that can be provided via xDSL transmission. These include distance learning, high graphic on-line commerce, video and audio entertainment, interactive advertising, news and other information services, advanced manufacturing processes, media production, remote security, public safety, telemedicine, utility management, advanced communication systems and transportation.

Customers do not care or need to know how the underlying broadband communication technologies function. They just care that technology works for whatever application they want to use. Broadband technologies, such as xDSL services, merely provide a system that enables applications to fulfill the needs or desires of end-users.

Much of the demand for broadband data access has come from the Internet. The Internet or the World Wide Web (WWW) has had a profound impact on our lives - both on a personal and business basis. The Internet's global collection of interconnected computer networks has become the medium of choice for dissemination of knowledge and connectivity for both individuals and corporations. The Internet is transforming the world into an "information society". This medium has created an awareness of many new information services, and these new information services are often best delivered via rapid speed data communication services.

Providing access to the Internet web is not an application, it is a service. This service standardizes the format of graphics and audio for distribution through data communications networks. Applications such as distance learning, on-line shopping, and news services use these standards to enable customers to obtain the benefits of the applications.

In 1999, the United States lead the "information society" with over 90 million Internet users with more than 33% of households with on-line web access. [1]. The Internet has transitioned from text based email and file transfer in the mid 1990's to digital audio broadcasting and animated advertising in the early 2000's. The demand for high bandwidth applications such as animation, video streaming, and high speed connections to corporate networks has lead to a multimegabit bandwidth race. Digital subscriber line (DSL) technologies are main contenders for this high bandwidth race.

In mid-2000, there were over 1,000,000 DSL customers in the United States [2]. Estimates show the market for DSL customers will reach 7.74 million residential customers and 1.83 million business lines, for a total of 9.5 million DSL lines deployed by 2003 [3].

High-speed connectivity is becoming a commodity. Service providers are adapting their networks and strategies to meet the demand for high bandwidth services and applications at reduced cost. Some service providers are attempting to capture high bandwidth customers by offering access to key information service providers (e.g. a specific Internet service or digital video provider).

The customer's key interest in high bandwidth service includes speeding up large file transfers, viewing high-resolution images and enabling multimedia applications such as streaming audio and video. The attempt of service providers to restrict access to certain information content providers is likely to decrease attention to developing cost effective, high-speed networks, and create opportunities for other high-speed network providers to offer access to other information content providers.

Early adopters of broadband services are more affluent. Of the broadband users in 1999, 21 percent of these households have annual incomes over $100,000. Broadband users were also twice as likely to be active online purchasers when compared to low speed users [4].

In a recent survey of online consumers, 80 percent stated that they would pay approximately $25.00 per month for broadband access alone, of which 26 percent said that they would pay $50.00 or $60.00 for a package of broadband enabled applications (that includes premium quality downloadable music or video files) on top of the cost of broadband access [5].

Many of these applications and services are increasingly being grouped according to type or nature of content, such as information or entertainment. It is also a common trend to have more and more of these tailored to individual tastes. They will be based on the user's own information, i.e. "my news", "my banking", "my investment portfolio", —- to make them absolutely specific and relevant to the user. Additionally every user will be able to create their own content including video, animation, still images and text where all the information is in digital transferable form.

It is predicted that there will not be a single "killer" application for broadband service because there are many high value broadband applications. Some of the grouping of services include: distance learning, high graphic on-line commerce, video and audio entertainment, interactive advertising, news and other information services, advanced manufacturing processes, media production, remote security, public safety, tele-medicine, utility management, advanced communication systems and transportation.

Distance Learning

Distance learning is the process of providing educational training to students at locations other than official learning centers (schools). Distance learning has been available for many years and is now used in elementary education (grades K-12), higher education (college), professional (industry), government training and military training. In the early years, distance learning was provided through the use of books and other printed materials and was commonly referred to as "correspondence courses".

Distance learning has evolved through the use of broadcast media (e.g. televisions) and moved onto individual or small group training through the availability of video based training (VBT) or computer based training (CBT). These systems have developed to interactive distance learning (IDL) as the computer allowed changes in the training.

Distance learning relies on communication systems (e.g. phone lines or mail) to connect students and teacher as an alternative to classroom training. Electronic learning (eLearning) is a form of distance learning that is becoming a viable option to traditional teaching methods and is poised for major growth over the next several years.

Through the ability of broadband video and interactive graphic technologies, students will be exposed to a far greater stimulus than in the traditional learning environment. Integrated sound, motion, image and text will all serve to create a rich new learning atmosphere and substantially increase student involvement in the learning process.

The rapid changing global economy is forcing industry professionals to continually update their skill sets. Adults are now changing their occupations several times in a lifetime as technologies and skill sets become outdated. This requires continual learning for adults. Adults between the ages of 35-45 are the fastest growing group of college learners [6]. To advance or consolidate their careers, over five million adults complete some form of distance learning each year in the United States [7]. This is one of the primary reasons why online learning is booming, especially among working adults with children. Distance learning via DSL connectivity allows adults to "attend" classes in the comfort of their living room or study, at their convenience.

Many of the online universities, including training and professional specialty course programs are catering to the rising demand from industry to deliver skill-development courses to the desktop at remote locations. These schools are offering Web-based professional certificates as well as associate and bachelor's degrees that are built around a solid core of business and computer classes. Companies rely on these certificates to ensure employees are qualified for their new jobs.

In 1999, most online classes did not require that students have the latest high-powered computer. However, they did require Internet access (via low speed analog modem). These distance-learning courses were provided using low resolution graphics or slow scan web video. As broadband services become more available and cost effective, it is predicted that distance-learning courses will evolve to use high resolution services such as high resolution video conferencing [8]. On-line distance learning courses can be accredited by regional accrediting agencies or via the Distance Education and Training Council.

Elementary (K-12) Education

By year-end 1998, approximately 89 percent of all public secondary and 76 percent elementary schools in the US were connected to the Internet. Since then, public schools have continued to make progress toward meeting the goal of connecting every school to the Internet by the year 2000. (In 1994 only 35 percent of public schools were connected to the Internet). In addition to having every school connected to the Information Superhighway, a second goal is to have every classroom, library, and media lab connected to the Internet. Schools are making great strides to achieve this and in 1998, 51 percent of instructional rooms in public schools were connected [9].

Connection speed is one of the key determining factors to what extent schools make use of the Internet. In 1998, higher speed connections using a dedicated line were used by 65 percent of public schools. Additionally, large schools with Internet access are more likely to connect using broadband access technology [10].

The economics of education systems limit the offering of specific courses to regions that have a density of students. To ensure that each student can be offered the same education opportunities, distance education can offer more courses to each student. Distance education allows students to interact with other students with similar interests and needs.

Higher Education

Since the Internet was pioneered at universities to facilitate information sharing, it's not surprising that an increasing number of them are creating Web-based universities. By 2002, 85 percent of two-year colleges (in 1999 there were a total

847 two-year colleges in US) are expected to be offering distance-learning courses. This is a 58 percent increase when compared to 1998. It is projected that over 80 percent of the four year colleges (in 1999 there were a total 1,472 four-year colleges & universities in the US) will be offering distance learning courses in 2002, up from 62 percent in 1998. Many of which will be Web-based. To put into perspective, there are 15 million full and part time college student in the US of which an estimated 90 percent are online, representing by far the most active single group on the Net. Moreover, in 1998, 21 percent of these students purchased $900 million in goods and services via on-line [11].

It is estimated that 93 percent of distance learning programs in American colleges and universities use e-mail and almost 60 percent use the e-mail in conjunction with the Web [12].

When distance education is offered, campus visits are not required for most programs. Learners register online each semester, and may take single courses for personal enrichment or opt to enter a degree program. Textbooks and class syllabi can be mailed to learners. Online classes run typically on a 16-week semester schedule, beginning and ending the same time as on-campus classes. Students read their textbook and visit online message boards weekly, posting class comments or questions whenever it is convenient for them. The back-and-forth commentary on the message boards simulates a classroom discussion. Midterm and final exams are usually taken under the watchful eye of an approved proctor at a local college, library, or human resources training center.

Figure 10.1 shows that the number of college students enrolled in distance learning courses in 1998, was approximately 710,000 and the number is estimated to reach 2.2 million in 2002; which will represent 15 percent of all higher education students [13].

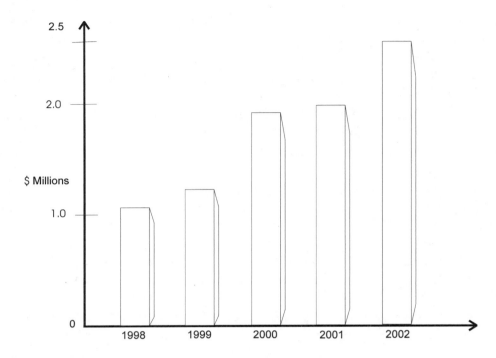

Figure 10.1, Growth of Distance Learning

Source: IDC (International Data Corporation); UT Austin Web Central

Professional

Professional education is developed and provided for companies, because it is necessary for companies to remain competitive. Technology and business processes are constantly changing. Training budgets range from 1% to 5% of a company's gross sales.

Government

In the United States in 1999, there were over 3 million government workers. The average government worker receives 1-2 weeks of training per year to use software and technology systems, standard processes, and develop leadership skills. This results in a requirement of over 5 million weeks of training.

Military

In 2000, there were over 5.8 million personnel in the United States military [14]. Military personnel continually require training to enable the use of high technology systems and equipment and develop leadership skills. Military provides a strong competitive advantage for any country. In the United States military, new entrants receive 1 to 6 months of technical training and active duty personnel receive 1-4 weeks of professional training each year.

Online Retail

Online retail involves the bartering and exchange of goods via a communication line to the Internet. In 1998, online retail in the United States exceeded $7.8 billion [15].

Online retail via the Internet web provides far greater product selection and information over the traditional brick and mortar retailers. There is a proliferation of automated shopping applications from business to consumer. These may be transaction based, but may also involve a wide range of products and services, such as purchasing books, CDs, flowers, and airline tickets. Additionally "intelligent personal search agents" will be able to get best prices, last minute deals, and help in suggesting the appropriate gift.

Broadband access will enhance online shopping. High-resolution catalogs, personalized animated modeling and video clips will increase the interest in online shopping. The number of online shoppers in the United States is projected to grow to 85 million, and the spending per shopper will double as well by 2003. Additionally, the duration of time between going online and making the first purchase is down to an average 4 months. In 1999, the median income for online households is 57 percent higher than that of the average American household — $58,000 vs. $37,005 respectively [16].

Figure 10.2 shows that the most popular e-retail categories in 1999 include: books, music, videos, apparel, health & beauty, food & beverage, airline tickets, computer hardware, hotel reservations and consumer electronics.

Online shopping provides tremendous advantages over brick and mortar retail store counterparts. Apart from the convenience of not having to drive to a store, the customer can usually specify preferred details of the product groups they are

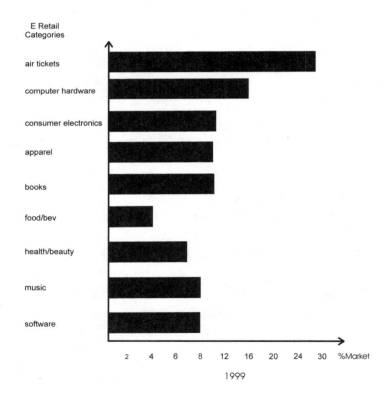

Figure 10.2, Division of E-Commerce Sales

Source: The National Retail Federation/ Forrester Research Online Retail Index, March 2000

looking for. Free online search engines such as Yahoo, Excite, and others are electronic directories that allow customers to rapidly find the companies that offer products that interest them.

Some of the key areas for online shopping in 1999 include travel (primarily air), software, music, health and beauty, food and beverage, books, apparel, consumer electronics, and computer hardware. Online shopping allows customers to define key attributes of the products they are looking for and to be almost instantly presented with selections that match their specific requests.

Travel

In 1999, US consumers booked $6.5 billion of leisure and unmanaged business travel online, almost triple the $2.2 billion booked in 1998, representing five percent of total US bookings in 1999. Online bookings are expected to increase significantly to 14 percent of total bookings by 2005 ($28 billion), with key segments including lodging, cruise, tour, and rental car products [17].

Software

Since the mid 1990's, companies have used the Internet to deliver software modules and programs. Until high speed broadband data transmission was available, software downloads were primarily limited to software repairs and device drivers. Broadband data transmission allows companies to deliver full multi-megabyte versions of software programs direct to consumers. Some of the popular options for software delivery via the Internet include time limited trial samples and limited versions of programs.

Music

Online shopping allows customers to easily preview content or details of a product such as tracks on music albums. In 1998, music industry revenue topped $13.5 Billion in the US, with online sales totaling $157 million, up 315 percent from 1997's figure of $37 million. The online sales of CDs, tapes and records is expected to grow to $2.6 billion (or 14 percent of total US music sales of $18.4 billion) in 2003.

The sales of digital downloads or distribution of music has been very limited due to bandwidth constraints, downloading a full-length CD, even in compressed form is a formidable challenge for the average user with a dial-up modem. With broadband access expanding, that the market for digital distribution of music should start taking off, and in 2002-2003 it is estimated that digital download of music revenue will be approximately $150 million. With DSL it will be possible to download entire music CDs in less than 2 minutes [18].

As an interim approach to music content delivery on the Internet, companies are offering digitally compressed music on the Internet in MP3 form. In 2000, a company MP3.com launched subscription music channels on the Net. For a monthly fee of less than $10, users have access to thousands of music tracks to choose from which will be available for listening.

Books

In 1999, over 50 percent of online shoppers indicate that they have purchased a book or CD online. The entertainment products category (books, video, and audio CDs) collectively, will grow to almost $9 billion in 2003 [19]. Online toy shopping is expected to reach $1.6 billion in 2003, up from $300 million in 1999 [20]. The leading book retailer online in 1999 was Amazon with over $1.2 billion in book and media sales [21].

Although the sale of books online is substantial, the need for high bandwidth links to assist with the sale of books is small. However, the availability of high bandwidth links will allow promotional video audio clips that will increase sales.

Apparel

The online apparel marketplace includes the sale of clothes, shoes, and personal luggage (e.g. handbags). The availability of high bandwidth access lines will allow for better imaging of apparel items. This may include 3 dimensional modeling and adaptation of product image products to the personal requirements (e.g. personal size). Items may be displayed with, or on, stored pictures of the customer.

Food and Beverage

Online food and beverage sales consist of coffee, gift fruit packs and almost all other edible goods. Creative online food and beverage marketers have advanced packaging technology to allow for the shipment of perishable goods to end consumers.

Food and beverage promoters use value added incentives such as high product quality and packaging (e.g. gift packs) for selling food and beverage products online. Improvements in food and beverage online marketing will likely include digital video clips of the benefits of using products.

Consumer Electronics

Selling consumer electronics equipment online primarily consist of product comparison of audio and photography equipment. Marketing consumer electronics online will benefit from high bandwidth marketing through displaying product features in use and by providing training for the use and comparison of electronics equipment.

Computer Hardware

During 1999, over 96% of online shopping involved order entry and less than 4% of the online shopping involved content delivery [22], primarily software products. In 1999, the primary online shopping content delivery was in the form of document (information content such as news services) or image delivery. As low cost broadband services become readily available, online content delivery will include movie rental, audio CD downloading, interactive gaming and more. It is expected that the percentage of online shopping revenue derived from content delivery will increase to over 8% by 2003.

Online Commerce

Online commerce (e-commerce) accounted for an estimated $57 billion in revenues in 1998 and analysts are forecasting that worldwide e-commerce will grow to $2.7 trillion by 2004, up from $740 billion in 2000. Accordingly e-commerce transactions in business-to-business will generate nearly $850 billion by 2004. Companies are expected to purchase on average 30 percent to 50 percent of goods and services online [23].

This surge can be attributed to the explosive growth on online user population, increased merchant activity, and increased consumer comfort with online transactions. E-commerce permits management of vital information such as customer profiles and data, orders, tracking and returns.

Online Trading

Online trading allows customers to research companies, identify business trends, receive live quotes, joint tip lists and stock forums/ chat rooms, and track your portfolio live.   Online trading allows customers to manage their personal investment transactions. The online brokerage assets are expected to quadruple from more than $750 million in 1999 to over $3 trillion in 2003 [24].

Online Banking

Online banking services make it possible for a customer to track checking and savings account statements, checkbook balancing, transfer funds and pay bills — all with built-in security from home or other location that is connected to the Internet. It's like having a bank teller available any hour of the day (and without the long line-ups).

There are approximately 1389 banks in the US with over 20 thousand bank branch locations in the United States and there were over 200,000 ATM (automated bank machines) as of 1998 [25]. Online banking reduces the need for customers to visit branch locations.

Although early computer banking services met with limited consumer acceptance, the awareness of the Internet and automated banking services has helped online banking rank higher in consumer importance.   This increasing level indicates consumers' acceptance of the online channel as a way to manage personal finances. Since 1998, there has been a significant increase in online banking features on bank sites. By 2000, over 90 percent of banks are expected to allow users to track account balances online. Additionally, as more bills become available through banks and more consumers adopt bill presentment as a standard format for bill payment, online management of personal finances will become increasingly important to consumers, and significant to financial services players. Online bill payment services on bank sites increased 47 percent from 1998 to 1999. Stock and bond trading capabilities on bank sites are also increasing [26]. Figure 10.3 shows that there is a significant advantage for banks to use online banking services.

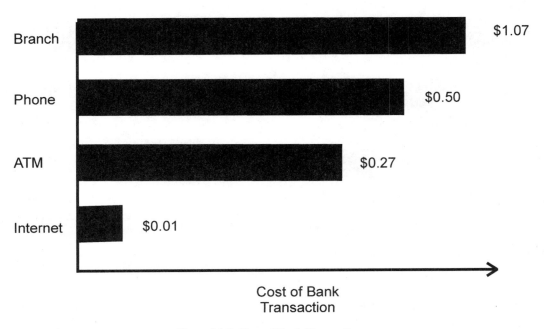

Figure 10.3, Cost of Bank Transactions

Source: Merrill Lynch

Bill Presentation and Payment

Bill presentment is the delivery of interactive electronic bills over the Internet. Bill presentment and payment offers the online consumer improved levels of ease, convenience, management and cost savings. Automatic bill presentation and payment options will compel consumers to demand online financial services and transactions. In 1998, there were approximately 15 billion recurring consumer paper bills were received by one hundred million US households. On an annual basis, the typical US household spends on average 24 hours on bill management, $46 on postage, and $144 on check-writing fees on 12 recurring monthly bills, growing to 15 million by 2002 [27]. By 2002, its estimated that some 60 percent of all online financial services households (defined as online households that online trade and online bank) will be using online bill presentation & payment, will be able to view on average 70% of their total bills, for a total of 1.4 billion bills presented to them online.

On-line billing presents significant cost savings for vendors that bill their customers. Each paper bill costs vendors between $1.50 and $3.50, whereas electronically presented bills will cost only $0.35 to $0.50 [28].

Online Entertainment

Online entertainment services include gambling, networked games, interactive toys, video and audio content delivery to the customer. The Internet is the world's biggest playground, and online entertainment is one of the fastest growing revenue areas on the Internet. Multiple research studies have shown that more than one-third of respondents say that they interact on Internet for a fun, leisure activity.

Gambling

Online gambling is the interactive process of allowing customers to wager money or credits in return for games that have standardized odds. Online gambling has the potential to be one of the largest interactive services as it taps into an existing traditional gambling. As of 2000, that gambling market was valued at over $900 billion world-wide. Customers with a credit card and an Internet connection are able to gamble. Casinos, lotteries and sports-books (horse & dog racing, boxing, team sports betting, etc.) are available anywhere the world.

While there are some issues about the legality of gambling online in the United States, the majority of online gamblers are U.S. citizens, it should be noted that most services are operating from places such as the Caribbean, Europe, Australia and South Africa. It projected that over $10 billion will be gambled online by 2002 as operators take advantage of the huge audience reach and cost savings of the Internet [29].

Networked Games

Since 1997, networked games have become a big opportunity on the Internet. Networked games allow users to play games against friends who are connected to the Internet. Almost any computer game that can be played by two or more people can be played online. It is estimated that by 2002, 60% of kids online (over 16 million) will be playing games and they will spend over $70.00 per year for game services. Adult will spend $140.00 per year for an estimated total of $622 million for online game services. High bandwidth services allow for substantially improved game services through streaming video and audio.

And as low cost home broadband services and equipment become more available, companies will create richer gaming applications. Major game vendors such as Nintendo, Sony, and Sega are beginning to enter the broadband market by selling game CDs and allowing potential customers to participate in online games for free. Alternatively there are some games that can only be played online, include: Ultima, Starseige, Quake Arena, and Unreal Tournament.

Customers pay a monthly access fee or pay-per-play to access online games. Networked games make it much easier for customers to find new opponents, or to find a partner to play at any time. Broadband high-speed data access will provide for much better 3 dimensional (3D) graphics. High-resolution 3D games usually require much higher data communication speed.

Interactive Toys

Interactive toys use communication technology to interact with other toys. Interactive toys have motors, sensors, infrared messaging and speech recognition technologies that respond to communication signals and originate messages. The responses may be in the form of some mechanical action or audio message.

Interactive toys have been available for many years. Some of the first interactive toys responded to signals that were sent via the television channel. These toys responded to colors or patterns within the television signal in the form of toy operation.

The interactivity of user interfaces are constantly improving. With high bandwidth connectivity to their home, these interactive toys, such as virtual pets, with artificial lives will become more alive offering simultaneous physical, verbal and PC-like interactive toys.

Movie Rental

Video content delivery will be one of the leading drivers of the online entertainment marketplace. Consumers have a voracious appetite for all types of media, particularly video (movie) content. In 1999, over 70 percent of US households rented an average of 1.3 videos per week [30]. The statistics for movie rentals confirm the preference of movie viewers to stay at home to view movie content. Since 1980, when VCRs first emerged as a means of watching full length motion pictures, the sales of pre-recorded rental and sell through video cassettes has grown over 66,000 percent to 1998 as compared to box office growth of 22 percent over the same period [31]. The video rental business is projected to top $7 billion in 2000 growing to $19 billion by 2004, and video sales reaching $20 billion [32].

Adult entertainment content ordering and delivery has been one of the leading categories of early Internet usage. As such adult entertainment was an early adopter of user interface augmentation, streaming video, and one click ordering. In 1998, pay-per-view and subscription adult entertainment accounted for about 40 percent of the US consumer paid online content market [33]. Adult entertainment, a multi-billion dollar industry should benefit from broadband access — who's consumers will no longer have to tolerate distorted images, business-card size, 10 frames per second videos.

Music Content

Although much of the content was not delivered via the Internet in the 1990's, much of the purchases of music came from online sales. The sales of CDs, tapes and records online is expected to grow to $2.6 billion (or 14 percent of total US music sales of $18.4 billion in 2003).

Until the late 1990's, much of the music content was primarily delivered to consumer's via CD and audio cassette sales. As broadband access becomes more cost effective, consumers are expected to shift from purchasing CD and cassette sales to online content delivery.

Interactive Movies

Interactive movies allow different scenes to be presented to a movie viewer dependent on the viewer's responses (e.g. via a keypad unit) or other source (media control center). Interactive movies offer a significant advantage for home viewers compared to movie-going viewers where as each home viewer can have a different experience. Interactive movies provide the viewer with an opportunity to feel more of a part in the movie viewing experience. With a multiplicity of windows, lobbies and screening rooms, information about new movies such as the plot, cast, crew and also cast interviews as well as viewer reviews may be dynamically selected.

Virtual Radio Stations

Virtual radio stations are digital audio sources that are connected to a network (typically the Internet). In 1999, there were over 2000 radio stations operating on the Internet. Virtual radio stations have a strong competitive advantage compared to standard radio broadcasts. Radio station web sites can do more than simply rebroadcast their on-air signals. They can provide photos of disc jockeys, contest prizes and their winners, act as current news centers for entertainment events and weather services.

Radio stations are taking aggressive steps in developing a new breed of web sites designed to offer fresh content and help the media outlets connect better with their target audience. This includes offering chat rooms, news updates and music reviews, and other social based services that make their web sites more appealing. Additionally, virtual radio stations can use their Web sites as research tools to determine listener preferences. The system serves up real-time information, providing details on the music being played by the radio station. Listeners then are asked to use the web site to vote on the song being played, giving station programmers instant feedback on listener tastes. The radio stations then talk up their web sites on-air, driving more usage of the Internet service. The Internet is having impacts on the radio stations ratings and revenues [34].

Virtual Television Stations

Virtual television stations distribute digital video and audio through the Internet to groups of viewers. With broadband DSL digital video access, the Internet will become a new avenue of distribution for broadcasters that hope to target previously unreachable office audiences.

Since 1999, there has been a growing public interest in interactive TV (iTV). This has been led by satellite and cable systems as they begin to deploy subscriber equipment and infrastructure capable of delivering a variety of interactive services. Some of the early interactive functions include: an electronic program guide (EPG) and parental control through channel-locking features, as well as supporting a type of one-way datacasting. Datacasting allows viewers to choose from limited, primarily text-based supplementary content.

Other virtual television features and functions may include game/ quiz show audience participation. These features and functions all presents new opportunities as well as challenges to programmers, advertisers, and providers of interactive services as they navigate through a maze of complex platform landscapes defined by a complicated mix of networks, set-top boxes, and software. Its projected that 35 percent of US households (over 25 million homes) will use some form of interactive TV services by the end of 2005 [35].

Virtual Books/ E-Books

Virtual books or electronic books (E-books) are books in digital form that can be displayed on computers or personal digital assistants (PDAs). PDAs are devices with bright monochrome LCDs designed for digitally encoded "print" content. Some of the devices come with leather covers, built-in modems, and color screens.

Since 1998, online publishing offered electronic books in postscript descriptor file (PDF) or Portable Document Format. A document that is converted into PDF looks identical to a print out version of the original document. PDF adds some functionality and also reduces the file size. (You cannot edit/ change PDF files). E-books offer book publishers a mechanism to control distribution if they're able to tie content to a specific device, $$$ electronic books, or e-books. In 1999, the total US book market was approximately $21 billion and the e-books market share was less than 1% [36]. This is due in part to poor display devices and the lack of compelling content and in part to the limitations of the user experience (which far outstrip those of digital music users). With the introduction of better display devices and more content available via the Internet, the marketplace for virtual books should dramatically increase. It is likely that E-book vendors will focus initially on vertical opportunity segments, such as books used in education, healthcare, and law.

Virtual Newspapers and Magazines

In 1998, the Internet emerged as a trusted news source with consumers gravitating towards it as a source for headline breaking news. Over 80 percent of survey respondents believe the Internet is as reliable as off-line media sources. With the proliferation of 24 hour cable news channels, and increasing online news services (AOL News Service is the most trafficked area in terms of unique accounts), the average daily newspaper readership falling to only 58 percent of the US population in 1997 from over 80 percent in 1964, and only 31 percent of 21 to 35 year olds reading the newspaper, traditional newspapers and broadcasters are looking to the Internet as an opportunity to reach the advertiser coveted, more affluent, younger demographic online audience [37].

Newspapers rarely duplicate themselves word for word online, but they often provide more than enough for the reader to live without the paper edition. When viewing online newspapers, readers are not restricted to a limited selection of local newspapers at nearby newsstands. They have access to virtually every newspaper on the planet (almost all have an online version). Additionally, the online versions of the newspaper are generally free and are available before the paper ones hit the stands. The online versions tend to offer expanded coverage into areas such a travel, entertainment and culture, exclusive content such as breaking news, live sports coverage, online shopping, opinion polls, and discussion groups. However, probably the best advantage of online newspapers is that they provide search and retrieve archives. With increased bandwidth access offered by DSL, virtual newspapers will be able to take advantage of video and audio to add value to the news services.

Electronic Photo Album

Since the mid 1990's, low cost digital cameras have been available that allow customers to capture and transmit digital photographs. Because digital cameras allow the customer to manipulate the digitized photos, they can be enhanced to remove red-eye, aligned, and unwanted areas can be cut out. These images can be used to create electronic postcards or greeting cards. Broadband data transmission allows customers to transmit and receive photos in seconds rather than several minutes.

Digital camera revenue is expected to surpass that of film cameras in 2000 for the first time ever, with $1.9 billion worth of digital cameras to be sold in the United States. Digital camera unit sales are expected to grow from 6.7 million in 2002 to over 42 million in 2005 [38]. Figure 10.4 shows a digital camera that

Figure 10.4, Digital Camera

Source: Hewlett Packard

is capable of capturing images in digital form and transferring these images to storage media (e.g. CD Rom) or transfer them via the Internet via email.

Business Applications

Key business applications that use broadband data transmission include video conferencing, remote corporate network connections, business kiosks, documentation control, customer care and field service support.

Video Conferencing

Video conferencing is the combination of dedicated audio, video, and communications networking technology for real time interaction. Companies can use video conferencing to reduce or eliminate travel while allowing employees to interact. Video conferencing facilities often combine various types of media. Products such as Microsoft's NetMeeting offers conference attendees at two (or more) locations real-time voice and video conferencing plus things like collaborative application sharing, document editing, background file transfer, and whiteboard to draw and paste on. Low cost high bandwidth data transmission of broadband systems will improve the quality and available video conferencing services. Figure 10.5 shows a digital camera that can be used for low cost video conferencing through the Internet.

Remote Corporate Network Connections

Remote corporate network connections allow company employees to access company networks and receive similar services (e.g. rapid file transfer) as they would experience if they were located (working) at the corporation. The rise in "Virtual Corporations" productivity improvements, facilities costs, environmental and regulatory issues such as anti-pollution rules and incentives to encourage workers telecommuting. These and life style changes towards more work-at-

Figure 10.5, Web Camera

Source: Logitech

home are major drivers causing business to push the envelope on network speed.

Business Kiosks

Business kiosks are the remote location of business retail centers. Business kiosks may be unmanned, or kiosks be operate with staff personnel at remote offices to provide connection to a head office. The use of business kiosks allows companies to expand their market territories without significant risk or capital investment.

Public Internet Kiosks are a type of payphone booth that contain a computer terminal that can access the Internet. For a nominal price, a customer can check their e-mail, last night's sports scores or surf the net. Most public Internet kiosks are scattered throughout public places such as airports, train stations, convention centers, hotels, office building lobbies, and shopping malls. These public Internet kiosks can be used as a media center for information services. They can be used as automated teller machines, travel service providers, ticket centers and other business services.

Corporations are using a virtual public network (VPN) business model that enables their employees to have publicly available Internet access as a viable way to pick up e-mail and check information. In 1998, there were approximately 10,000 kiosks in the US, and the number raising to more than 100,000 by 2002. The typical cost of a kiosk is in the $35,000 to $55,000 range, in addition to monthly space rental fees [39].

Documentation Management

Documentation management includes the capture, storing, organizing and coordinating access to large amounts of text and image information. This information may be stored at one or more locations and the information may be accessed or transferred to display devices (terminals), printers or other repositories (for long term storage).

Documentation management allows manuals, procedures, specifications and other vital information to be instantly accessible by authorized employees. Documentation management can save a company a considerable amount in printing reproduction costs, as all documentation is digital rather than paper.

Field Service

Field service personnel interact with clients or equipment in the field. These personnel have had limited access to company materials. Using broadband communications systems, field service personnel can access documents (e.g. company catalogs and service manuals), example procedures (e.g. video clips), capture information (e.g. using a digital camera to record an insurance claim), and to assist in the repair of equipment (e.g. connect systems for remote diagnostics). Figure 10.6 shows a wireless personal digital assistant (PDA) that allows a field service representative to access various forms of media.

Figure 10.6, Personal Digital Assistant

Source: Nokia

Customer Care

Customer care is the process of answering customer questions about a company's products or services. It is estimated that over 65% of the cost of customer support comes from simple product and billing questions [40].

The cost of customer service is greatly reduced and customer satisfaction is dramatically improved as customers and suppliers are able to satisfy their information need via the Internet. Additionally, the information gathered from customer browsing (areas regularly visited) will allow companies to promote similar products and services.

Media Production

Media production involves the coordination of artistic content creation into various electronic medium forms prior to physical production.

Image and Video Production

Images and Video can be captured in electronic form and transferred to other locations. Because of the large file size of high-resolution images, media transfer has primarily been in the form sending high-density disks or video tapes.

Printing Press

Printing presses transfer electronic media information (usually in the form of an image file) onto a print medium (usually paper for books, newspapers or magazines). Since the early 1990's, printing presses have been converting their mechanical reproduction processes from mechanical to electronic (digital presses). In either case, printing presses require film or another medium that uses high resolution images that are usually created by computers. High bandwidth systems can do this faster.

Telemedicine

Telemedicine is the providing of medical services with the assistance of telecommunications. Telemedicine does not completely replace medical expertise, telemedicine is critical to providing quality and efficient health care services.

Telemedicine is a rapidly growing part of the medical information management market and is one of the largest and fastest growing segments of the healthcare device industry. The expected revenue by the year-end 2000 is $21 billion. In the USA more than 60% of federal telemedicine projects were initiated in the last 2 years. The concept of telemedicine captures much of what is developing in terms of technology implementations, especially if it is combined with the growth of the Internet and World Wide Web (WWW) [41].

If planned and managed properly, telemedicine will increase the quality of care in a given geographic area and allow hospitals to dramatically increase their market reach. Telemedicine is also the most effective means of managing a capitated patient population, especially when that population includes rural residents.

The telemedicine networks encompass computer, video, and telecommunications technologies; each with its own role to play in the acquisition, transport, and display of medical information.

With the increasing integration of Web-based information retrieval and exchange, and movement toward the computer-based patient record, telemedicine is migrating to the desktop.

Medical information gathered and input, either manually or electronically, creates a telemedicine file using software that integrates the data, voice, images, or video. That file can be transmitted to a similar workstation at a medical center, physician's office, or other site equipped to manage the telemedicine information request. Rapidly transporting image data and diagnoses between clinicians and medical doctors can add substantially to patient care.

Some of the advanced telemedicine applications include Telecardiology, Teleradiology and Telepsychiatry. Telecardiology services incorporate transmission of ECG data, echocardiograms, heart sounds and murmurs, and cardiology images, and can be performed in both store-and-forward and interactive media. Teleradiology is the most widely adopted of all telemedicine applications. Clinical radiology requires prompt, near real-time transmission of still-frame images, but may also demand live or full-motion video image communication and display. Telepsychiatry allows psychiatric care to be conducted at a distance and telepsychiatry can provide care more frequently to patients in outlying areas.

Remote Clinical Facilities

With high-resolution desktop video conferencing capabilities, patients can meet with clinical technicians at remote facilities instead of traveling long distances to meet with medical experts. The clinical technician coordinates communications with medical experts via video conferencing consultations.

The zoom capabilities of video conferencing systems are particularly prized in remote Telepsychiatric applications. This capability makes it easier to assess the effects of medications or other diseases that can cause involuntary body movements. The use of video conferencing reduces the anxiety of patients as the ability to zoom makes medical observations less intrusive.

Diagnosis and treatment of skin problems is another promising telemedicine application. A video dermascope with a polarizing light source can visualize a dermatologic lesion. The dermascope provides an industry-standard video signal for input into the computer workstation via video-capture board. Dermatology biopsies normally viewed under a microscope use the same capturing devices as the pathology application.

Telemedicine video conferencing facilities allow hospital-based physicians to view patient wounds from a live video image. The traditional method requires visiting nurses to take Polaroid photographs of wounds and forward them to physicians for review. From the snapshot, the physician assesses how the wound is progressing and determines whether changes in medication or treatment are needed. Using the telemedicine DSL system, visiting nurses dial the physician, forward the image in real-time, and facilitate interaction between patients and hospital-based providers. Images can be captured and stored in an electronic medical record. The technology can help reduce the cost of continuing inappropriate therapy and shorten the time between data collection and decision.

Manufacturing

Information systems have long been used in manufacturing processes to control production and ensure the quality of products. Production monitoring, engineering and research and development processes all will benefit from low cost data communication systems.

Production Monitoring

Production monitoring is the process of using data devices or sensors (e.g. video cameras and keypads) that transfer information via communications lines to keep records of physical production. The Internet and other communication networks are moving onto the factory floor providing companies with an inexpensive means to link workers and the machines they operate to remote repositories of information. Distant managers can watch what's going on, literally, from wherever they are, as sensors, tiny web cameras and web displays are being built directly into equipment deployed on assembly lines and by using Internet technologies to speed products to market. Browser technology is becoming the standard interface and has training advantages and saves software development time and trouble. Software that integrates Internet technologies into factory operations is a small but a growing portion of a $4.8 billion market in 1999 for prepackaged manufacturing software, which itself is growing 14.2 percent a year [42].

Advanced Communications Services

Advanced communications services include public telephone network bypass, Internet telephony, electronic mail, and high-speed network interconnections. Broadband technology offers the ability to provide multiple voice channels in addition to a high-speed data channel.

Telephone Network Bypass

Telephone network bypass is the connection of a customers voice circuits (e.g. PBX system) to public telephone network systems (e.g. long distance services) without using the local telephone companies switching system. Telephone network bypass permits competing voice services companies (e.g. MCI and Sprint) to avoid paying high tariffs that are required by local telephone companies. Telephone network bypass routes only the data portion of the DSL copper access line to an alternative voice switching or routing network.

The data portion of each DSL line can provide 12 (SDSL) to over 800 (VDSL) standard voice circuits (DS0s). More digital voice circuits could be provided through the use of digital voice signal compression (8:1 or more!). These digital voice communication circuits can be connected to other voice communications systems (packet voice networks) providing a bypass of local telephone companies. In the mid 1990's, cable companies began offering telephone service over cable networks. In 1999, Cox cable company stole 100,000 local telephone customers from Bell South and US West [43].

Internet Telephony

Internet telephony is the transferring of voice communication service via the Internet packet data network. The Internet has been used as an alternative to the telephone network since the mid 1990's. Because of the limited amount of hardware and software along with marginal voice quality, Internet Telephony had primarily been used to reduce the cost of calling long distance. Internet Telephony usually allows a customer to place the call through a personal computer using a standard sound card, speakers and a microphone. Many sound cards provide for duplex transmission that permits Internet telephony users to conduct a regular conversation like an ordinary telephone.

Unlike the telephone network that uses circuit switched technology, the Internet uses packet switching technology. The delays for packet switching are random and the present version of the Internet does not have a quality of service (QoS) structure to control the maximum delay. A new version of Internet packet switching is being developed that allows for QoS control (Internet version 6). This is likely to dramatically increase the amount of Internet Telephony service throughout the world.

Electronic Mail (email)

Electronic mail (email) is the transferring of information messages via an electronic communications system. Initial versions of email could send short text messages of 1 to 3 pages. Email technology has evolved (standardized) to allow file attachments and new versions of email (such as Flash technology) send animation or video clips as email messages.

E-mail messaging is probably the best single reason for connecting to the Internet, and has been arguably one of the greatest contributors to the growth of the Internet. There were over 400 million email account users in 1998, and the number of e-mail accounts is expected to top 1 billion by the end of 2000. E-mail messaging is quickly becoming a replacement for the post, fax and even voice telephone. E-mail is a great improvement on the postal system. It revolutionizes the way and the amount we communicate. E-mail permits us to send a message to anyone (or even copy an entire address book) anywhere in the world instantaneously. Email overcomes time, distance and even some culture barriers.

E-mail messaging has been the leading application ("killer application") among online users age 18 and younger. E-mail is used by greater than 40 percent of online kids under age 13, and almost 60 percent of online kids between ages 13 and 18. A large proportion of older kids spend their time online communicating with others (via email, chat, and instant messaging).

High Speed Network Interconnection

The initial development effort for DSL technologies was to allow for several voice communication circuits to share a single copper wire pair (pair gain services). The high speed data transmission of broadband DSL technologies are now being used to allow companies to interconnect corporate networks and telephone companies to interconnect switching systems.

References:

1. Jupiter Communications.
2. Interview, online analyst expert, 2 August 2000.
3. New Paradigm Resources and Forrester Research
4. Jupiter Communications Consumer Survey
5. Jupiter Communications 1999
6. Interview, education expert 29 July 2000.
7. Ibid.
8. Interview, industry expert, 6 May 2000.
9. US Department of Education/ National Center for Education Statistics
10. US Department of Education
11. e-Marketer; Student Monitor LLC
12. The Survey of Distance Learning Programs in Higher Education, 1999 edi-

tion

13. IDC (International Data Corporation); UT Austin Web Central

14. Personal Interview, Tara Ramos, Industry Expert, 14 July 2000.

15. Jupiter Communications

16. Activemedia 2000; e-Marketer

17. Jupiter Communications

18. Jupiter Communications; The Recording Industry Association of America/ RIAA

19. Ibid.

20. Ibid.

21. 10k, Amazon, Dec 1999.

22. Interview, industry expert, 6 May 2000.

23. Yankee Group, 2000

24. Jupiter Communications.

25. Midwest Research of Cleveland; ABA (American Bank Association "Statistical Informaiton on the Financial Services Industry, 8th edition").

26. ABA, Jupiter WebTrack, Merrill Lynch

27. Jupiter Communication

28. Jupiter Communications; Morgan Stanley

29. Datamonitor

30. Video Store Magazine, January 1999

31. Motion Picture Association

32. Paul Kagan Associates

33. Jupiter Communications

34. www.electricvillage.com

35. Jupiter Communications.

36. Interview, publishing distribution industry expert, 21 April 2000.

37. Newspaper Association of America; Pew Research Center.

38. InfoTrends report.

39. Summit Report.

40. Interview, Steve Kellogg, 6 May 2000.

41. Telemedicine Information Exchange (TIE).

42. IDC (International Data Corp.

43. Maribel Lopez, "Beyond Broadband," Forrester research, 2000.

Appendices

Appendix I

Acronyms and Abbreviations

A/D - Analog to Digital
ABR - Available Bit Rate
ACK - Acknowledge
ADC - Analog to digital converter
ADSL - Asymmetric Digital Subscriber Line
AIN - Advanced Intelligent Network
AM - (1) Amplitude modulation; (2) ante meridian, before noon.
ANSI - American National Standards Institute
AP - Adjunct Processor, Access Point
API - Application-Programming Interface
ARPANET - Advanced Research Projects Agency Network

ASIC - Application Specific Integrated Circuit
ATM - Asynchronous Transfer Mode
ATM25 - Asynchronous Transfer Mode 25 Mbps
ATU-C - ADSL Transmission Unit-Central Office
ATU-R - ADSL Transmission Unit-Remote Location
AWG - American Wire Gauge
B-ISDN - Broadband Integrated Services Digital Network
BER - Bit Error Rate
BOC - Bell Operating Company
bps - bits per second
BRI - Basic Rate Interference

CAP - Competitive Access Provider, Carrierless Amplitude and Phase

CATV - (1-) Cable Television; (2-) Community Antenna TV

CBR - Constant Bit Rate

CCBS - Completion of Calls to Busy Subscriber , Customer Care and Billing System

CCH - Common Control Channel

CCITT – (1)-International Telegraph and Telephone Consultative Committee (succeeded by ITU-T); (2)-Consulatative Committee on International Telegraphy and Telephony

CDR - Call Detail Recording

CDV - Call Delay Variability

CEBus - Consumer Electronics Bus

CELP - Code Excited Linear Predictive Coding

Centrex - Central Exchange

CEPT - European Conference of Posts and Telecommunications (standards activities succeeded by ETSI), Conférence Européenne (des Administrations) des Postes et des Télécommunications

CLEC - Competitive Local Exchange Carrier

CLID - Calling Line Identification

CLIR - Calling Line Identification Restriction

COAX - Coaxial Cable

CODEC - Coder-Decoder

COST - ETSI (previously CEPT) Council on Science and Technology, European Telecommunications Standards Institute

CPE - (1-) Cellular Provider Equipment; (2-) Customer Premises Equipment

CQM - Channel Quality Measurement

CRC - Cyclic Redundancy Code/Check

CSU - Channel Service Unit

CUG - Closed User Group

D-WDM - Dense Wavelength Division Multiplexing

DAML - Digital Added Main Line

dB - Decibel

DBS - Direct Broadcast Satellite

DES - Data Encryption Standard

DHCP - Dynamic Host Configuration Protocol

DISA - Direct Inward System Access

DLC - Digital Loop Carrier

DMT - Discrete Multi-Tone

DNS - Domain Name Server

DOC - (Canadian) Department of Communications

DS1 - Digital Signal (level 1)

DS3 - Digital Signal (level 3)

DSI - Digital (or dynamic) Speech Interpolation

DSL - Digital Subscriber Line

DSLAM - Digital Subscriber Line Access Multiplexer

DSMA - Digital Sense Multiple Access

DSP - Digital Signal Processing

DSU - Digital Service Unit

DTE - Data Terminal Equipment

DTMF - Dual-Tone Multifrequency (signaling)

DTV - Digital Television

DTX - Discontinuous Transmission

DUN - Dial-Up Network

DVI - Digital Video Interactive

E911 - Enhanced 911

ECMA - European Computer Manufacturers Association

EIA - Electronics Industries Association

EMAIL - Electronic Mail

EMI - Electromagnetic Interference

EO - End Office

EOC - Embedded Operations Channel

ETSI - European Telecommunications Standards Institute

FCC - Federal Communications Commission

FDD - Frequency Division Duplex

FDDI - Fiber Distributed Data Interface

FDM - Frequency Division Multiplexing

FDX - Full Duplex

FEC - Forward Error Correction

FER - (1-) Frame Error Rate (2-) Frame Erasure Rate

FEX - Foreign Exchange

FEXT - Far End Cross Talk

FITL - Fiber In The Loop

FM - Frequency Modulation

FOD - Fax-on-Demand

FPLMTS - Future Public Land Mobile Telephone System

fps - Frames Per Second

FRAD - Frame Relay Assembler/Disassembler

FSAN - Full Services Access Network

FSK - Frequency Shift Keying

FTP - File Transfer Protocol

FTTC - Fiber to the Curb

FTTH - Fiber to the Home

FTTN - Fiber to the Neighborhood

FX - Foreign Exchange

GEO – (1)-Geostationary Earth Orbit; (2)-Geosynchronous Earth Orbiting

GP - Guard Period

GPS - Global Positioning System

GTT - Global Title Translation

GUI - Graphic User Interface

HDSL2 - High Bit Rate Digital Subscribe Line 2nd Generation

HDTV - High-Definition TV

HDX - Half Duplex

HFC - Hybrid Fiber Coaxial

HOMEPNA - Home Phone line Networking Alliance

HTML - Hyper-Text Markup Language

IDLC - Integrated Loop Digital Carrier

IDT - Integrated Digital Terminal

IEEE - Institute of Electrical and Electronics Engineers

IETF - Internet Engineering Task

Force

ILEC - Incumbent Local Exchange Carrier

IN – (1)-Interrogating Node; (2)-Intelligent Network

IP – (1)-Intelligent Peripheral; (2)-Internet Protocol

IPC - ISDN to POTS Converter

IPR - Intellectual Property Rights

IPSec - Internet Protocol Security

IRTF - Internet Research Task Force

IS - Interim Standard

ISDN - Integrated Services Digital Network

ISI - Inter-Symbol Interference

ISO - International Standards Organization

ISP - Internet Service Provider

ISUP - ISDN User Part

ITU - International Telecommunication Union

IVHS - Intelligent Vehicle Highway System

IVR - Interactive Voice Response

IWF - InterWorking Function

IXC - Interexchange Carrier

JPEG - Joint Photographic Experts Group

KSU - Key Service Unit

KTS - Key Telephone System

L2TP - Layer 2 Tunneling Protocol

LAN - Local Area Network

LAP - Link Access Protocol

LATA - Local Access and Transport Area

LCR - Least Cost Routing

LDS - Local Digital Switch

LEC - Local Exchange Carrier

LED - Light Emitting Diode

LEO - Low Earth Orbit

LMDS - Local Multipoint Distribution Service

MAN - Metropolitan Area Network

MBONE - Multicast Backbone

MDF - Main Distribution Frame

MEO - Medium/Middle Earth Orbit

MFJ - Modified Final Judgment

MIME - Multipurpose Internet Mail Extension

MIPS - Million Instructions Per Second

MMDS - Multichannel Multipoint Distribution Service

MMI - Man Machine Interface

MOD - Music-on-Demand

MODEM - Modulator-Demodulator

MoU - Memorandum of Understanding

MPEG - Motion Picture Experts Group

MSO - Multiple System Operator

MTBF - Mean Time Between Failures

MUX - Multiplexing

NANP - North American Numbering Plan

NASTD - National Association of State Telecommunications Directors

NAT - Network Address Translation

NBS - National Bureau of Standards

NEXT - Near End Cross Talk

NIC – (1)-Network Independent Clocking; (2)-Network Information Center (Internet Registry); (3)-Network Interface Card

NID - Network Interface Device

NOC - Network Operations Center

NPRM - Notice of Proposed Rulemaking

NRZ - Non-Return to Zero

NT - Network Termination

NTIA - National Telecommunications and Information Administration

NTU - Network Termination Unit

NVOD - Near Video on Demand

OA&M - Operations, Administration and Maintenance

OMC - Operations and Maintenance Center.

ONU - Optical Network Unit

OSI - Open System Interconnection/Integration

OSS - Operator Services System or Operational Support System, operation subsystem

PABX - Private Automatic Branch Exchange

PAD - Packet Assembler/Disassembler

PAM - Pulse Amplitude Modulation

PBX - Private (automatic) Branch Exchange

PC - Personal Computer

PCI - Protocol Capability Indicator, Peripheral Component Interconnect

PDA - Personal Digital Assistant

PDN - Public Data Networks, Premises Distribution Network

PDU - Protocol Data Unit

PIN - (1) Personal Identification Number (2) Positive-Intrinsic Negative (photo-diode)

PM – (1)-Phase Modulation; (2)-also post meridian, afternoon; (3)-Physical Medium

POP - Point of Presence

POTS - Plain Old Telephone Service

PP - Point-to-Point

PPDN - Public Packet Data Network

PPP - Point-to-Point Protocol

PRI - Primary Rate Interface

PSC - Public Service Commission

PSDN - Public Switched Digital Network

PSDS - Public Switched Digital Service

PSK - Phase Shift Keying

PSTN - Public Switched Telephone Network

PTN - Public Telephone Network

PTT - (1) Postal Telephone and Telegraph (2) Push-to-Talk

PUC - Public Utilities Commission

QAM - Quadrature Amplitude Modulation

QBONE - Quality of Service Backbone

QOS - Quality Of Service

RACE - Research (and Development) of Advanced Communication (Technologies) in Europe

RADSL - Rate Adaptive Digital Subscriber Line

RBOC - Regional Bell Operating Company

RDT - Remote Digital Terminal

ROM - Read Only Memory

RP - Radio Port

S/N - Signal to Noise Ratio

SAP - Service Access Point

SCP - Service/Signal Control Point

SCTE - Society of Cable and Television Engineers

SDH - Synchronous Digital Hierarchy

SDSL - Symmetrical Digital Subscriber Line

SIM - Subscriber identity module

siu - Subscriber Interface Unit

SLIP - Serial Line Interface Protocol

SMDS - Switched Multimegabit Data Service

SMS - Service Management System or Short Messaging Service

SMTP - Simple Mail Transport Protocol

SNA - Systems Network Architecture

SNMP - System Network Management Protocol

SNR - Signal to Noise Ratio

SONET - Synchronous Optical Network

SP - Signaling Point

SS7 - Signaling System 7

SSL - Secure Socket layer

SSP - Service Switching Point

STP - (1-) Signal Transfer Point; (2-) Shielded Twisted Pair; (3-)Spanning Tree Protocol

T-Carrier - Trunk Carrier

TAPI - Telephony Application Programming Interface

TCAP - Transaction Capabilities Application Part

TCP - Transmission Control Protocol

TCP/IP - Transmission Control Protocol/Internet protocol

TDD - Time Division Duplex

TDM - Time Division Multiplexing

TDMA - Time Division Multiple Access

TE - Terminal Equipment

TIA - Telecommunications Industry Association

TMN - Telecommunication Management Network

TSAPI - Telephony Services Application Programming Interface

TSI - Time Slot Interchange

UBR - Unspecified Bit Rate

UNI – (1)-Universal Network Interface; (2)-User Network Interface

URL - Universal Service Locator

USB - Universal Serial Bus

USTA - United States Telephone Association

UTP - Unshielded Twisted Pair

UWAG - Universal ADSL Working Group

VAD - Voice Activity Detection

VBR - Variable Bit Rate

VC - Virtual Channel

VDSL - Very High Data Rate Subscriber Line

VM - Voice Mail

Vmail - Video Mail

VMSC - Visited MSC, Voice Mail System

VO-CODER - Voice Coder

VOD - Video on Demand

VODSL - Voice Over DSL

VOIP - Voice Over Internet Protocol

VPN - Virtual Private Network

VRU - Voice Response Unit

VSELP - Vector-Sum Excited Linear Predictive Coding

WAN - Wide Area Network

WDM - Wavelength Division Multiplexing

WER - Word Error Rate

WLL - Wireless Local Loop

WOTS - Wireless Office Telephone/Telecommunication system

WPBX - Wireless PBX

WRC - World Radio Conference

WWW - World Wide Web

Appendix II
Glossary

100 BaseT - A data communications system primarily used for computer networks based on the Ethernet IEEE standard. This system can be used on coaxial cable or twisted pair and it has a data rate of lOOMbps.

10BASET - A data communications system primarily used for computer networks based on the Ethernet IEEE standard. The 10BaseT system can be provided on coaxial cable or twisted pair and it has a data rate of lO Mbps.

2-wire circuit - A communication circuit that uses one pair of wires for both directions of transmission.

2500 telephone - A standard analog telephone device. This device usually provides plain old telephone service (POTS). This type of telephone is often referred to as a 2500 series telephone.

2B1Q (two binary, one quarternary) - A line code (modulation and signaling structure) that transmits two binary bits of data at one time using a multi-level code. This code was initially used by the Integrated Services Digital Network (ISDN) standard for basic rate service (2 64 kbps + 1 16 kbps channels). 2B1Q is used by some DSL technologies including IDSL, HDSL and SDSL.

3B2T line code - A baseband line code (modulation and signaling structure) that transmits three bits for every two ternary, three level symbols.

4-wire Circuit - A communication circuit that uses separate pairs of 2 pairs of wires for each direction of transmission.

4B3T line code - A baseband line code (modulation and signaling structure) that transmits four bits for every three levels of symbols.

Abilene - A high-speed communication network used in the next generation Internet2 system. Abilene is the backbone (interconnection) network that allows multimedia services through Internet2.

Access Line - The physical link (typically a copper wire or fiber) between a customer and a communications system (typically a central office) that allows a customer to access local and toll switched networks. Access lines may include a subscriber loop, a drop line, inside wiring, and a jack.

Access Point (AP) - Typically, a point that is readily accessible to customers for access to a wireless or wired system. Also called a Radio Port.

Acknowledgement message (Ack) - A message that is responded by a communications device to confirm a message or portion of information has been successfully received. If a communications device is supposed to send ack messages back to the originator and an acknowledgment message is not received, the system will typically re-send the message. See also negative acknowledgment message (Nack).

ADSL Forum - A forum that was started in 1994 to assist manufacturers and service providers with the marketing and development of ADSL products and services. The ADSL forum has been renamed the DSL forum.

ADSL modem - An electronics assembly device that modulates and demodulates (MoDem) asymetric digital subscriber line (ADSL) signals. ADSL signals are usually transmitted on a twisted pair of copper wires.

An ADSL modem may be in the form of an internal computer card (e.g. PCI card) or an external device (Ethernet adapter). Most ADSL modems have the ability to change their data transfer rates based on the settings that are programmed by the DSL service provider and as a result of the quality of the communication line (e.g. amount of distortion).

ADSL Transmission Unit - Remote (ATU-R) - An advanced modem that provides for asynchronous digital subscriber line (ADSL) multi-megabit data rates over unshielded twisted pair (UTP) of copper wires. The ATU-R is usually located at a customer's premises.

The ATU-R can be in various configurations including and internal computer modem (PCI bus), external modem that connects to the Universal Serial Bus (USB), or a bridge device that converts ADSL signals to an 10BastT or 100BaseT Ethernet form.

ADSL Transmission Unit - Central Office (ATU-C) - An advanced modem that provides for asynchronous digital subscriber line (ADSL) multi-megabit data rates over unshielded twisted pair (UTP) of copper wires. The ATU-C is usually located at a central switching office or at a remote distribution node (e.g. RTC). The ATU-C is essentially a mirror image of an ATU-R.

ADSL-Lite - A limited version of the standard ADSL transmission system. This limited version of ADSL allows for a simpler filter installation that can often be performed by the end user. The limitation of ADSL-Lite is a reduced data transmission rate of 1 Mbps instead of a maximum rate of 8 Mbps.

Advanced Intelligent Network (AIN) - A telecommunications network this is capable of providing advanced services through the use of centralized databases that can provide call processing and routing. AIN systems are capable of providing enhanced services that can be used on wireless and wired networks.

Advice of charge (AOC) - The ability of a telecommunications system to advise of the actual costs of telephone calls either prior or after the calls are made. For some systems, (such as a mobile phone system) the AOC feature is delivered by short message service.

Alert tones - The types of alert tones that are available to indicate a particular status of a telecommunications event. An example of an alert tone is a sound that alerts the user that a new short message has been received.

Alternate Mark Inversion (AMI) – A modulation code that is used by telephone companies for digital voice and data communication. AMI binary transmission uses return to zero (RZ) coding in an alternating bipolar scheme. The AMI system uses with logical zeros that correspond to 0 V, and logical ones alternating between +3 V and -3 V to indicate a one state. This process does allow self-synchronization is possible with this approach when a limited number of zeros is transmitted.

Alternative access provider - A telecommunications service that provides an access connection between the end customer and a telecommunications network. This provider is a different company than the established LEC or PTT company.

Always On - A connection to a communications network (such as the Internet) that appears always on to the customer. Although always on connections appear as a dedicated connection to the end user (no need to initiate a dial up sequence), the connection may be temporary and automatically re-established each time the user accesses the network.

American Wire Gauge (AWG) - A measurement system that provides the diameter of conductors (typically copper wire). The larger the thickness of the wire (higher the gauge), the lower the AWG number and the better of the ability of the line to transmit electrical signals. Many local telephone access loops use 24 AWG or 26 AWG copper lines.

Analog to Digital Converter (ADC) - A signal converter that changes a continuously varying signal (analog) into a digital value. A typical conversion process includes an initial filtering process to remove extremely high and low frequencies that could confuse the digital converter. A periodic sampling section that at fixed intervals locks in the instantaneous analog signal voltage, and a converter that changes the sampled voltage into its equivalent digital number or pulses.

Analog video - Analog video contains a rapidly changing signal (analog) that represents the luminance and color information of a video picture. Sending a video picture involves the creation and transfer of a sequence of individual still pictures called frames. Each frame is divided into horizontal and vertical lines. To create a single frame picture on a television set, the frame is drawn line by line. The process of drawing these lines on the screen is called scanning. The frames are drawn to the screen in two separate scans. The first scan draws half of the picture and the second scan draws between the lines of the first scan. This scanning method is called interlacing. Each line is divided into pixels that are the smallest possible parts of the picture. The number of pixels that can be displayed determines the resolution (quality) of the video signal. The video signal television picture into three parts: the picture brightness (luminance), the color (chrominance), and the audio.

Applications Program Interface (API) - A software program that is dedicated for the interfacing of specific types of hardware or software programs. API programs often allow end users to define the operation or sequence of tasks for complicated systems. Examples of API include Telephony Services API (TSAPI) and Java Telephony API (JTAPI).

ARPAnet - A computer network that was developed by the Advanced Research Projects Agency of the U.S. Department of Defense. ARPAnet was the predecessor to the Internet. The objective of ARPAnet was to allow continuous communications in the event portions of the network were disabled (possibly due to military or nuclear weapon attack).

Asymmetric Digital Subscriber Line (ADSL) - A communication system that transfers both analog and digital information on a copper wire pair. The analog information can be a standard POTS or ISDN signal. The maximum downstream digital transmission rate (data rate to the end user) can vary from 1.5 Mbps to 9 Mbps downstream and the maximum upstream digital transmission rate (from the customer to the network) varies from 16 kbps to approximately 800 kbps. The data transmission rate varies depending on distance, line distortion and settings from the ADSL service provider.

Asymptotic coding gain - A processing gain of a coding system that can be obtained when the signal to noise ratio (SNR) approaches infinity.

Asynchronous - A signal which does not have synchronization with some other reference signal. The communications on an asynchronous channel is not sequential and may appear random in nature.

Asynchronous Transfer Mode (ATM) - A packet data and switching technique that transfers information by using fixed length 53 byte cells. The ATM system uses high speed transmission (155 Mbps) and is a connection based system. When an ATM circuit is established, a patch through multiple switches is setup and remains in place until the connection is completed. ATM service was developed to allow one communication medium (high-speed packet data) to provide for voice, data and video service.

As of the 1990's, ATM has become a standard for high-speed digital backbone networks. ATM networks are widely used by large telecommunications service providers to interconnect their network parts (e.g. DSLAMs and Routers). ATM aggregators operate networks that consolidate data traffic from multiple feeders (such as DSL lines and ISP links) to transport different types of media (voice, data and video).

asynchronous transfer mode 25 Mbps (ATM25) - A 25 Mbps version of ATM. The ATM25 standard was developed primarily for corporate networks. However, the QoS advantages of ATM and customer needs for switched services for digital video and Internet access has stimulated interest in ATM for the DSL industry.

ATM Cell - A 53 byte packet of data (called a "cell") that is used in an ATM network. An ATM cell is usually divided into a 5 byte header and 48 byte payload. The ATM header is primarily used for local connection routing information to the next switching point.

Availability - A measurement that indicates the connection status or a commitment to provide a minimum amount of connection status of a network during a period of time. Availability may be measured by a connection time or by a minimum performance measurement. (e.g. at a minimum data transfer rate)

available bit rate (ABR) - A communications service category that specifies the maximum available bit transfer rate (ABR) transferred through a telephone network..

Backbone - A communication system that connects several network equipment together. A backbone system is usually a high-speed communications network such as ATM or FDDI.

bandwidth - The information bandpass capability of a communication channel as measured in Hertz. The bandwidth is the difference between the upper and lower frequency limit of the communications channel.

baseband channel - A information content (channel) that is used to modulate or encode a transmission medium. When used with radio signals, the high frequency component is called the broadband channel.

Basic Rate Interface (BRI) - The standard basic interface that is used in the integrated services digital network (ISDN) system. The BRI interface provides up to 144 kbps of information that is divided into two 64 kbps channels (voice or data) and one 16 kbps control channel (data). The 64 kbps channels are referred to as the B channels and the 16 kbps channel is called the D channel.

bearer services - A service of providing the communications path that carries user data. Bearer services do not process or store any of the user data.

best effort - A level of service in a communications system that doesn't have a guaranteed level of quality of service (QOS). Best effort services include DSL and cable modem access to the Internet.

Bit - The smallest part of a digital signal, typically called a data bit. A bit typically can assume two levels; either a zero (0) or a one (1).

bridge - A data communication device that adapts the line levels and protocols between two different types of networks. A bridge is sometimes called a gateway.

Bridge tap - A bridge tap is an extension to a communication line that is used to attach two (or more) end points (user access lines) to a central office. Bridge taps provide connection options to the telephone company on connecting different communication lines to a central office without having to install new pairs of wires each time a customer requests a new telephone line.

Bridged Tap - An communication line (tap) that is connected to another communication line between a receiver and a transmitter. The bridged tap appears as a stub or side branch on the main line. Bridged taps allow communication lines to have other possible termination points (possibly to allow connection to different customers in the future). At normal audio signal frequencies (300Hz to 3300Hz), a bridged tap does not significantly affect the electrical transmission characteristics. However, at high frequencies (such as those used in xDSL technologies), bridged taps can distort the transmission of electrical signals (commonly called a shunting affect).

broadband - A term that is used to describe high-speed data communications. Communication systems that have a data transmission rate above 1 Mbps are typically called Broadband systems.

Broadband Communications - Voice, data, and/or video communications at rates greater than wideband communications rates (1.544 Mbit/s).

broadcast fax - A process or service that broadcasts a fax message to a list of pre-defined recipients.

brownout - A reduction in the servicing of customers as a result of the demand for service exceeding the service processing capability of the service provider's equipment or staff. Brownout usually occurs during a peak period. Because service access attempts by customers increase during a brownout period (customers repeatedly attempt to get service), service providers may discontinue services to groups of customers during brownout.

browser - A software program or module (called a client) that is used to convert information that is available on the Web portion of the Internet into forms usable by a person (text, graphics and sound).

bundling services - The combining of different services into one service offering so the customer can communicate with one company for several different services. An example of bundling is the combination of cellular, PCS, local and long distance services as one service package.

burst - A short transmission of information (data).

cable modem - A communication device that Modulates and demodulates (MoDem) data signals to and from a cable television system. Cable modems select and decode high data-rate signals on the cable television system (CATV) into digital signals that are designated for a specific user.

There are two generations of cable modems; First Generation one-way cable modems transmit high speed data to all the users into a portion of a cable network and return low speed data through telephone lines or via a shared channel on the CATV system. First generation cable modems used asymmetrical data transmission where the data transfer rate in the downstream direction was typically much higher than the data transfer in the upstream direction. The typical gross (system) downstream data rates ranged up to 30 Mbps and gross upstream data rates typically range up to 2 Mbps. Because 500 to 2000 users typically share the gross data transfer rate on a cable system, cable modems also have the requirement to divide the high-speed digital signals into low speed connections for each user. The average data rates for a first generation cable modem user rage up to 720 kbps. Until the late 1990's, most cable modems used first generation technology.

Second generation cable modems offered much data transmission rates in both downstream and upstream directions. Second generation cable television systems use two-way fiber optic cable for the head end and feeder distribution systems. This allows a much higher data transmission rate and many more channels available for each cable modem. As of the year 2000, approximately 35% of the total cable lines in the United States had already been converted to HFC technology.

Cable Television (CATV) - A television distribution system that uses a network of cables to deliver multiple video and audio channels. CATV systems typically have 50 or more video channels. In the late 1990's, many cable systems started converting to digital transmission using fiber optic cable and digital signal compression.

call barring - A telecommunications network feature, which restricts the origination or delivery of calls to a device operating in its system. Call barring can restrict some or all parts of outgoing or incoming calls. When all calls are restricted (may be activated when the customer does not pay their bill), no calls may be made from or answered by the telephone device. Outgoing call restriction may limit the calls to emergency only, no international calls or local calls only. Restrictions on incoming calls may allow calls from specific people in a closed user group (CUG) or non-toll calls (such as collect calls).

call detail record (CDR) - A call record that holds the origination and destination address of a call, time of day the call was connected and duration of the call.

call hold - A feature that allows a user to temporarily hold and incoming call, typically to use other features such as transfer or to originate a 3rd party call. During the call hold period, the caller may hear silence or music depending on the network or telephone feature.

Caller ID - A service that provides a receiving telephone device with the phone number of the originating caller, which can be displayed to the person prior to receiving the call. Some caller ID services can also provide name information. Caller ID information is typically transferred as a high frequency data signal between the ring cycles of the alerting tone.

calling line identification restriction (CLIR) - The ability of a caller to block the delivery of their telephone number from being displayed on a caller identification device. The CLIR may be on a per call basis or per line (continuous) block. CLIR provides the ability for the caller to remain anonymous to the called person.

carrier - (1 - radio) A single radio signal, when transmitted, that may be modulated by another wave containing information. A radio carrier can be modulated by various amplitude, frequency and phase changes. (2 - dropout) A short time period where the carrier signal is lost. (3 - frequency) The frequency of the radio carrier signal. (4 - level) The radio energy (power) of a carrier signal, typically expressed in decibels in relation to some nominal (reference) level. (5 - telecommunication service) A company engaged in transferring electrical signals or messages for hire.

Carrierless Amplitude and Phase (CAP) - Carrierless amplitude and phase (CAP) modulation is very similar to QAM modulation. The difference is the continuous shifting of phase (or signal mix) of the carrier signal level. CAP modulation was designed to help to reduce the effects of crosstalk and to simplify the signal processing of modulated signal. CAP transmits data signals on a single high bandwidth modulated carrier.

cell - (1 - cellular system) A radio coverage area associated with a fixed-location cellular or PCS radio tower that is interconnected with other cells to provide radio coverage to a larger geographic area. The term cell is often visualized as a hexagon as a relative building block depicting the ability of a cellular system to continually split so the system capacity can continually increase as new customers are added to the system. (2 - battery) A primary (disposable) or secondary (rechargeable) unit that stores energy for the supply to electrical or electronic equipment. (3- fuel) An electrochemical cell that produces electricity from the chemical energy of a fuel and an oxidant.

channel - (1 - general) The smallest subdivision of a circuit that provides a single type of communication service. (2 - broadcasting) A portion of the radio frequency spectrum assigned to a particular broadcasting station. (3- signal path) A transmission path between two or more termination points. The term channel can refer to a 1-way or 2-way path. (4 - video effects) A digital effects processing path for video. (5) A single unidirectional or bi-directional path for transmitting or receiving, or both, of electrical or electromagnetic signals.

channel capacity - The amount of data or channel transmission capability of a communication channel.

channel coding - Channel coding is a process where one or more control and user data signals are combined with error protected or error correction information. After a sequence of digital data bits has been produced by a digital speech code or by other digital signal sources, these digital bits are processed to create a sequence of new bit patterns that are ready for transmission. This processing typically includes the addition of error detection and error protection bits along with rearranging of bit order for transmission.

channel spacing - The spacing in a radio frequency band (in Hertz) between adjacent radio carrier signals. It is measured from the center of one channel to the center of the next adjacent channel.

churn - (1 - turnover) A term which denotes loss of customers. (2 - special services) The number of special-service circuit additions that are predicted to be exactly balanced by an equal number of special-service circuit disconnects.

circuit - (1 - communication) Any communication path through which any information can be transferred. (2 - electronics) A combination of electrical processing components that perform a process (such as signal amplification) or function (clock display processor).

circuit switched data - The continuous transfer of data between two points. To establish a circuit switched data connection, the address is sent first and a connection (may be a virtual connection) path is established. After this path is setup, data is continually transferred using this path until the path is disconnected by request from the sender or receiver of data.

circuit switching - A process of connecting two points in a communications network where the path (switching points) through the network remains fixed during the operation of a communications circuit. While a circuit switched connection is in operation, the capacity of the circuit remains constant regardless of the amount of content (e.g. voice or data signal) that is transferred during the circuit connection.

Class 4 - A voice communications switching system used to interconnect local telephone switching systems. The class 4 switching system was one level above the class 5 end office switching system. Class 4 switches are also known as "Tandem Switches."

Class 5 - A classification of a switching that is used by local telephone service provider. A class 5 switch is the last point in the network prior to the customer. Class 5 switches usually can handle from 10,000 to 100,000 customers. A class 5 switch is also called a "Central Office" (CO).

closed user group (CUG) - (1- access restriction) A group of directory numbers sharing an access restriction such that any directory number can reach others in the group but cannot access outside numbers. (2- cellular system) Advanced features such as 4-digit dialing authorized for a closed group of users of the service. (3 - X25 protocol) In the X.25 packet-switching protocol , a facility indicating a virtual grouping of terminals that can communicate only with other members of that group. The feature can be extended to a closed user group with outgoing access, or a closed user group with incoming access.

coaxial cable (coax) - A cable is composed of an inner conductor that is completely surrounded by the outer conductor. Coax cable was designed to transfer high-frequency signals with a reduced amount of signal leakage.

CODEC - An acronym for COder/DECoder, a device that turns analog signals into digital signals for transmission and then decodes the digital signals back into its original analog form at the receiver.

coding - (1 - digital) A process of changing digital bits to include error protection bits and/or signaling bits prior to the sending or storing of the information. (2 - software) The process of writing instructions or commands for software programs.

Competitive Access Providers (CAPs) - A telecommunications service provider that offers competing services to an established (incumbent) telephone service provider. CAPs typically compete with a local exchange carrier (LEC). CAPs can provide service by reselling local service from the LEC.

Competitive Local Exchange Carrier (CLEC) - A telephone service company that provides local telephone service that competes with the incumbent local exchange carrier (ILEC).

Competive Local Exchange Carrier (CLEC) - A telecommunications service provider that is not an incumbent local exchange company (LEC) that offers telephone access services.

component analog video (CAV) – Signals, which represent the luminance and color information of a video picture. Each signal contains an analog voltage that varies with picture content. CAV also is referred to as analog component.

Conference of European Postal and Telecommunications Administrations (CEPT) - A European standards body composed of national telecommunications administrators and official carriers. Also known as the Conference Europeenne des Administrations des Postes et des Telecommunications.

constant bit rate (CBR) - A class of telecommunications service that provides an end user with constant bit data transfer rate. CBR service is often used when real time data transfer rate is required such as for voice service.

Constellation - A method of displaying multiple information parts of an signal. A constellation may display phase and amplitude on an x-y axis. Using the final parameter of time where the signal has a cycle of one rotation around the x-y axis, points on the constellation may display information elements (logic 1's and 0's).

Consumer Electronics Bus (CEBus) - A communications transmission system that transfers data on local (typically residential) power lines. The standard for CEBus is EIA IS-60. CEBus works similar to an Ethernet data network system.

contention - A condition which exists when two or more devices attempt to transmit at the same time using a shared channel.

Convolutional Coding - A error correction process, which uses the input data to create a continuous flow of error protected bits. As these bits are input to the convolutional coder, an increased number of bits is produced. k is the length of the code. Convolutional coding is often used in transmission systems that often experience burst errors such as wireless systems.

copper cross connect - A copper cross connect system allows access lines (copper lines) to be connected to several different xDSL modems. There are two key reasons to use a copper cross connect system. The first reason is to allow a copper wire access line to be connected to different digital subscriber line modems. This could be because the customer may upgrade to a new type of modem (e.g. ADSL to VDSL) or if a xDSL modem fails, a spare DSL modem could be connected to the customer's line. The second reason is to allow the access line to be connected to a DSL modem only when a connection is required. This would allow a DSL service provider to install lesser number of modems in a system than they have customers for.

crosstalk - The undesired leakage of a signal from one communications channel to another.

customer - Any individual, partnership, association, joint-stock company, trust, corporation, governmental entity, or any other entity that orders access services from an exchange, interexchange, or international, carrier. A customer also may be called a subscriber.

customer care - The process of communicating with the customer regarding their account, service feature selection, billing rates and invoicing.

customer care and billing system (CCBS) - A system that provides customer account tracking, service feature selection, billing rates, invoicing and details.

customer premises equipment (CPE) - All telecommunications terminal equipment located on the customer's premises, including telephone sets, private branch exchanges (PBXs), data terminals, and customer-owned coin-operated telephones.

cyberspace - A term that is commonly used to describe an interconnected pubic network (such as the Web) that has tools available that allow users to find and retrieve data from the network.

Cyclic Redundancy Check (CRC) - An error detection and/or correction method that is used to determine if a series of data bits were received correctly during transmission. To setup a CRC error checking process, the original bits of data are supplied to a CRC generator. The CRC generator uses a specific mathematical formula to create a new group of data bits called the CRC check sum (bits). The CRC check bits are typically appended to the data bits that are being sent. The receiver of the message compares the result of the CRC generator on the receiver end to determine if the bits were received without error. In some cases, CRC check bits can be used to help correct some bits that were received in error during transmission..

demarcation point - The physical and electrical boundary between an end user's telecommunication equipment and the telecommunications network.

department of communications (DOC) - A government agency of countries throughout the world that sets policies and rules regarding telecommunications within their country. In the United States, the FCC is the equivalent of the DOC in some other countries.

dial-up network (DUN) - A software portion of Microsoft Windows 95,98,NT,and 2000 that allows the user to connect the computer to a data network (such as the Internet). Because DUN is actually a process of establishing and maintaining a communications session, DUN is sometimes used for establishing connections on "always-on" circuits (such as DSL).

digital - Digital refers to a signal or category of electronic devices that represent information by discrete signal levels that change at predetermined intervals. Digital signals typically vary in two levels; on (logic 1) and off (logic 0).

digital added main line (DAML) - A local loop access line system that uses ISDN digital transmission to provide two communication circuits on a standard copper wire pair. DAML differs from the standard ISDN basic rate interface (BRI) as the D signaling channel is not included. The D channel is simply not used. The line control (e.g. off-hook) is sensed by DAML modem. The modem creates the line control signaling messages. DAML allows a telephone service provider to add more lines on existing copper access lines without the need to add ISDN software upgrades to their switching offices.

digital broadcasting - The process of transmitting the same digital data signal to all users that are connected to the digital broadcast network. Digital broadcast signals may be encoded in a way that only some of the users may be capable of decoding digitally broadcast messages (e.g. a specific pay-per-view movie channel).

digital compression - Digital compression is a process that uses a computing device (such as a digital signal processor) to analyzes a digital signal and create a new data signal that represents of the original signal using a lesser number of digital bits. Digital compression allows more information to be transmitted on a communication channel.

Digital compression devices use mathematical formulas and code book tables to compress the data. Mathematical formula transform the original signal into its characteristic parts such as frequency and amplitude. Code book tables contain blocks of high occurrence information (such as particular tones used in fax machines). When transmitting digital information that has been compressed, only the parameters (such as the frequency, amplitude and code book word) are sent on the transmission channel. When compressed digital information is received, a decoder reverses the compression process to produce the same (or similar) initial signal.

digital loop carrier (DLC) - A high efficiently digital transmission system that uses existing distribution cabling systems to transfer digital information between the telephone system (central office) and a users telephone and/or computer equipment. A DLC system usually includes a high-speed digital line (e.g. T1) from a central office and a remote digital terminal (RDT). The RDT converts the high-speed digital line to low speed lines (analog or digital) for routing to the end customers.

digital power line - Digital power line is a term that refers to the sending of digital information through electric power lines.

Digital Signal 1 (DS1) - A standard digital transmission line that is divided into twenty-four 64 kbps channels (commonly called voice channels). A DS1 line is divided into a 193 bit frames and transmitted at 1.544 Mbps. DS1 signals can be transmitted in an unframed form where frames are 192 bits at a data transmission rate of 1.536 Mbps.

Digital Signal 3 (DS3) - A standard digital transmission line that is divided into twenty-eight DS1 (T1) channels. The gross transmission rate for a DS3 channel is 44.736 Mbps. A single DS3 provide for 672 standard (64 kbps) voice channels.

digital signal processor (DSP) - An electronics circuit (typically an integrated circuit) that is designed to process signals through the use of embedded microprocessor instructions.

digital subscriber line (DSL) - A two-wire, full-duplex transmission system that transports user data between a customer's premises and a digital switching system or remote terminal at 144 kbps.

Digital Subscriber Line Access Multiplexer (DSLAM) - An electronic device that usually holds several digital subscriber line (DSL) modems that communicate between a telephone network and a end customers DSL modem via a copper wire access line. The DSLAM concentrates multiple digital access lines onto a backbone network for distribution to other data networks (e.g. Internet).

digital television (DTV) - A process or system that transmits video images through the use of digital transmission. The digital transmission is divided into channels for digital video and audio. These digital channels are usually compressed. Video compression commonly uses one of the motion picture experts group (MPEG) standards to reduce the data transmission rate by a factor of 200:1.

digital video - Digital video is a sequence of picture signals (frames) that are represented by binary data (bits) that describe a finite set of color and luminance levels. Sending a digital video picture involves the conversion of a scanned image to digital information that is transferred to a digital video receiver. The digital information contains characteristics of the video signal and the position of the image (bit location) that will be displayed.

digital video interactive (DVI) - The type of product that combines digital video and audio and allows the computer user to control the operation of the media display. DVI is a registered trademark of Intel.

direct broadcast satellite (DBS) - A satellite with enough range and power to be received by small dish antennas suitable consumer home use. DBS can be sent to both direct individual homes as well as received by communities by means of retransmission over a small TV station or cable TV system. In the late 1990's, the DBS marketplace become a formidable competitor to the traditional cable industry. DBS systems provider digital-quality pictures and have the potential to offer high speed interactive services. By using digital compression technology, DBS systems can offer a greater number of channels than analog cable systems to both PCs and TVs. DBS systems can also be customized to provide unique services for limited video on demand (VOD), near video-on-demand (NVOD) and interactive pay-per-view channels.

Direct Inward Service Access (DISA) - The process of how incoming calls are handled to a telephone system.

discontinuous transmission (DTx) - The ability of a communications system to inhibit transmission when no or reduced activity is present on a communications channel. DTx is often used in mobile telephone systems to consesve battery life.

Discrete Multi-Tone (DMT) - A data communications process that transfers a high speed data communication channel by dividing it into several narrow sub-channels and sending them independently through frequency divided channels. When the sub-channels are received, the low speed parts are recombined to create the original high-speed data transmission signal.

The advantage of sending several sub-channels is the ability to independently adjust the transmission levels of each sub-channel signal. Because the frequency response of the line can vary and distortions can occur on specific frequencies (where only a few sub-channels may be affected). DMT is used in xDSL systems as it adapts well to the hostile environment of copper wire transmission.

distance learning - Distance learning is the process of providing educational training to students at locations other than official learning centers (schools). Distance learning has been available for many years and is now used in elementary education (grades K-12), higher education (college), professional (industry), government training and military training. In the early years, distance learning was provided through the use of books and other printed materials and was commonly referred to as "correspondence courses."

Distance learning has evolved through the use of broadcast media (e.g. televisions) and moved onto individual or small group training through the availability of video based training (VBT) or computer based training (CBT). These systems have developed to interactive distance learning (IDL) as the computer allowed changes in the training.

distribution cable - A cable or cabling system that is used to transfer signals from a central location (e.g. a central office or the head end of a CATV system). to end customers.

domain name server (DNS) - The data processing device that translates text and numeric names for an Internet addresses. A DNS uses a distributed database containing addresses of other DNS servers that may contain the Internet address.

downlink - (1- Satellite) The portion of a communication link used for transmission of signals from a satellite to a mobile or fixed receiver. (2- cellular system) The radio link from the base station to the mobile station.

downstream - The direction of transmission, usually from a network to an end customer.

drop wire - A drop wire is the wire or pairs of wires that are connected between a customer's premises and a nearby network line. Although the first drop wires were connected from a telephone pole to a building, drop wires can be buried or aerial.

DSL bridge - A device that translates the protocol between a DSL modem and a DSL network. A DSL only translates the protocol and does not assign a separate address to the end user.

DSL Forum - A forum that was started in 1994 to assist manufacturers and service providers with the marketing and development of DSL products and services. The DSL forum was previously called the ADSL forum.

duplex - The transmission of voice or data signals that allows simultaneous 2-way conversations.

dynamic host configuration protocol (DHCP) - A process that dynamically assigns an Internet Protocol (IP) address from a server to clients on an as needed basis. The IP addresses are owned or controlled by the server and are stored in a pool of available addresses. When the DHCP server senses a client needs an IP address (e.g. when a computer boots up in a network), it assigned one of the IP addresses available in the pool.

dynamic IP addressing - A process of assigning and Internet Protocol address to a client (usually and end user's computer) on an as needed basis. Dynamic addressing is used to provide an enhanced level of security (no predefined address to use for hackers) and to conserve on the number of IP addresses required by a server. Also see DHCP.

e-commerce - A shopping medium that uses a telecommunications network to present products and process orders. Also called "virtual mall."

E1 - A communication line that was developed by European standards that multiplexes thirty voice channels and two control channels onto a single communication line. The E1 line uses 256 bit frames and transmitted at 2.048 Mbps.

Eb/No - The ratio of bit energy to a noise signal.

echo - A type of transmission impairment in which a sage signal is reflected back to the originating source. In the transmission, the reflected signal often is attenuated and delayed, resulting in an echo.

echo canceller - A signal processing device or circuit that reduces the effects of echo signals. Echo canceling is performed by calculating an estimate of the expected echo signal, by subtracting this estimate from the signal in which the echo appears, analyzing if the subtraction produced a signal without an echo and removing the echo signal if the analysis shows there was an echo. Echo cancellers are essential for communication systems that have long signal processing delays such as long distance voice, satellite and digital mobile telephony lines.

echo canceling - A process of extracting an original transmitted signal from the received signal that contains one or more delayed (usually from multi-path reflections) copies of the original signal. Echo canceling is performed via advanced signal analysis and filtering.

Electronic Industries Association (EIA) - A trade association that develops standards for electronic components and systems and represents manufacturers of electronic systems and parts.

Electronic mail (Email) - A process of sending messages in electronic form. These messages are usually in text form. However, they can also include images and video clips.

embedded operations channel (EOC) - A communications channel that is designed to be part of a communications circuit. The EOC allows for commands (usually system step, change and test commands) to be transferred without the need to interfere with an established communications link it is paired with.

encryption - Encryption is a process of a protecting voice or data information from being obtained by unauthorized users. Encryption involves the use of a data processing algorithm (formula program) that uses one or more secret keys that both the sender and receiver of the information use to encrypt and decrypt the information. Without the encryption algorithm and key(s), unauthorized listeners cannot decode the message.

Enhanced 911 (E911) - A emergency telephone calling system that provides an emergency dispatcher with the address and number of the telephone when a user initiates a call for help. The E911 system has the capability of indicating the contact information for the local police, fire, and ambulance agencies that are within a customers calling area.

Equalization - A processes which modifies the receiver parameters to compensate for changing radio frequency conditions. Primarily used to compensate for multi-path propagation (see section RF channel).

error correction - A process of using some data bits that are transmitted along with the data message to help correct bits that were received in error due to distorted radio transmission. Error correction is made possible by sending bits that have a relationship to the data that is contained in the desired data block or message.

error protection - The process of adding information to a data signal (typically by sending additional data bits) that permits a receiver of information to detect and/or correct for errors that may have occurred during data transmission.

Ethernet - A packet-switching transmission protocol that is primarily used in local area networks (LANs). Ethernet is a registered trademark of Xerox Corporation.

European Telecommunications Standards Institute (ETSI) - An organization that assists with the standards-making process in Europe. They work with other international standards bodies, including the International Standards Organization (ISO), in coordinating like activities.

far end crosstalk (FEXT) - The leakage of signal that is coupled to a nearby cable or electronics circuit (called crosstalk) where the unwanted signal is received on the far end (remote end) of the cable.

fast Ethernet - The standard Ethernet protocol that transfers data at 100 Mbps capacity. Fast Ethernet is commonly called 100BaseT.

fault tolerance - The ability of a network or sub-system to continue to operate in the event of a hardware or software failure. Fault tolerant systems are typically able to identify the fault and replace the failed component or sub-system with another equipment.

FCC type approval - A approval from the FCC that identifies the radio equipment manufactured has passed tests certifying it meets the minimum FCC requirements for that type of radio equipment. Most radio devices must meet several FCC specification requirements to receive FCC type approval. Companies typically use an independent testing lab to certify that equipment meets FCC requirements.

Federal Communications Commission (FCC) - A federal agency of the United States that establishes and enforces laws and regulations regarding interstate radio and wired communications services. The agency was established by the Communications Act of 1934.

feeder cable - Feeder cables are used to connects a central office or head end facility to distribution cables.

firewall - A data filtering device that is installed between a computer server or data communication device and a public network (e.g. the Internet). A firewall continuously looks for data patterns that indicate authorized use or unwanted communications to the server. Firewalls vary in the amount of buffering and filtering they are capable of providing. An ideal (perfect) firewall is called a "brick wall firewall."

Firmware - Firmware is software program instructions that are stored in a hardware device that performs data manipulation (e.g. device operation) and signal processing (e.g. signal modulation and filtering) functions. Firmware is stored in memory chips that may or may not be changeable after the product is manufactured. In some cases, firmware may be upgraded after the product is produced to allow performance improvements or to fix operational difficulties.

follow-me phone service - A service that allows calls to be routed to a customer's choice of forwarding phone numbers. Follow-me service may be automatic (e.g. when a cellular telephone automatically registers with a visited systems) or manually set (e.g. when a customer calls in with a hotel phone number where they will be temporarily located).

Forward Error Correction (FEC) - A system of error protection used in communication systems (typically wireless) that allows a receiving device to use the received data bits to detect and correct errors that occur during the transmission of data. FEC is performed by sending additional bits or by forming a relationship between the transmitted bits to provide redundancy of information transmission.

frame - (1 - general) A portion of a communication channel that repeats and typically constructed of groups of fields or time slots. (2 - equipment) A electronic rack that is used to interconnect and hold electronics assemblies. (3- video) Information that completes a full picture in a video system. (4 - data) In data transmission, the sequence of contiguous BITs bracketed by and including beginning and ending flag sequences. Unit of data of the data link layer.

frame relay - A packet-switching technology provides dynamic bandwidth assignment. Frame relay systems are a simple bearer (transport only) technology and do not offer advanced error protection or retransmission. Frame relay were developed in the 1980s as a result of improved digital network transmission quality that reduced the need for error protection. Frame relay systems offer dynamic data transmission rates through the use of varying frame sizes.

frequency - A measure of the number of complete signal cycles of a waveform that occur within a given length of time, typically specified in cycles per second (hertz).

Frequency Division Multiplex (FDM) - Frequency division multiple is used to divide a frequency bandwidth into several smaller bandwidth frequency channels. Each of these smaller channels is used for one communications channel.

frequency hopping - A radio transmission process where a message or voice communications is sent on a radio channel that regularly changes frequency (hops) according to a predetermined code. The receiver of the message or voice information must also receive on the same frequencies using the same frequency hopping sequence.

Frequency hopping was first used for military electronic countermeasures. Because radio communication occurs only for brief periods on a radio channel and the frequency hop locations are only known to authorized receivers of the information, frequency hopping signals are difficult to detect or monitor.

Frequency Shift Keying (FSK) - A form of frequency modulation in which the modulating signal shifts the output frequency between predetermined values.

Full services access network (FSAN) forum - A forum that was established in 1995 to help identify technologies and network architectures that can cost effectively provide narrowband and broadband telecommunications services.

future public land mobile telephone system (FPLMTS) - The third generation mobile telephone requirements as specified by the international telecommunications union (ITU). This system combines may of the features used in wired telephone systems with wireless networks.

G.dmt - G.dmt is an ITU standards for asynchronous digital subscriber line (ADSL). G.dmt permits data transmission rates of up to 8 Mbps downstream and 1.54 Mbps upstream.

G.lite - The limited version of asynchronous digital subscriber line (ADSL) technology that eliminates or reduces the need the installation of a splitter at the end customers location. The standard allows up to 1.5 Mps downstream and 384 Kbps upstream.

gateway - A communications device or that transforms data that is received from one network into a format that can be used by a different network. A gateway usually has more intelligence (processing function) than a bridge as it can adjust the protocols and timing between two dissimilar computer systems or data networks. A gateway can also be a router when its key function is to switch data between network points.

Global Positioning System (GPS) - Global Positioning System is a navigation system that uses satellites to act as reference points for the calculation of navigational position. GPS is used extensively by the military and aircraft GPS chipsets are now being incorporated into wireless devices, including phones, PDA's, as well as automotive applications.

Graphic User Interface (GUI) - The use of graphics (typically on a computer monitor display) to interface the output or requested input of a software application with a user. The use of buttons, icons and dynamically changing windows are typical examples of a GUI.

guard band - A portion of a resource (frequency or time) that is dedicated to the protection of a communication channel from interference due to radio signal energy or time overlap of signals. While guard bands protect a desired communication channel from interference, the guard band also uses part of the valuable resource (frequency bandwidth or time period) for this protection.

guard period (GP) - A time period that is a portion of a burst period where no radio transmission can occur. The guard period is used to protect adjacent burst from transmission overlap due to propagation time from the mobile radio to the base station.

Half Duplex - The ability to transfer voice or data information in either direction between communications devices but not at the same time. The information is transmitted on one frequency and received on another frequency.

High bit rate Digital Subscribe Line (HDSL) - An all digital transmission technology that is used on 2 or 3 pairs of copper wires that can deliver T1 or E1 data transmission speeds. HDSL is a symmetrical service.

High bit rate Digital Subscribe Line 2 (HDSL2) - A second generation of HDSL that offers several enhancements to HDSL data transmission. These improvements include the ability to transfer T1 or E1 data transmission rates over a single twisted-pair local loop instead of the two or three pairs of copper wire required for standard HDSL.

Home Phoneline Networking Alliance (HomePNA) association - An association that assists in the development of phoneline networking standards that allow for data to be distributed to devices that are connected to telephone wiring in a home or business.

HomeRF - An industry working group that is assisting in the development of a local area RF communications that permits consumer devices such as computers, printers and fax machines to communicate with each other.

hub - An communication device that connects several devices in a network (often data communication devices). A hub generally is a simple device that distributes data messages to multiple receivers. However, hubs can include switching functional and multi-point routing connection and other advanced system control functions.

hunting - A telephone call-handling feature that causes a transferred call to "hunt" through a predetermined group of telephones numbers until finds an available ("non busy") line.

Hybrid Fiber Coax (HFC) - The hybrid fiber coax (HFC) system is an advanced CATV transmission systems that uses fiber optic cable for the head end and feeder distribution system and coax for the customers end connection. HFC are the 2nd generation of CATV systems. They offer high-speed backbone data interconnection lines (the fiber portion) to interconnect end user video and data equipment. Many cable system operators anticipating deregulation and in preparation for competition began to upgrade their systems to Hybrid Fiber Coax (HFC) systems in the early 1990's. As of late 2000, over 35% of the total cable lines in the United States had already been converted to HFC technology.

hypertext markup language (HTML) - A text based communications language that allows formatting and item selection features to be transferred independent of the type of computer system. HTML is primarily used for internet communication.

IEEE 802.11 - IEEE 802.11 is an industry standard for wireless network communication. It usually operates in the 2.4 GHz spectrum and permits data transmission speeds from 1Mbps to 11 Mbps.

IEEE 802.3 - The industry standard for the Ethernet local area network (LAN) protocol. The 802.3 standard allows 10 Mbps or 100 Mbps data transmission rates. Because Ethernet can use CDMA/CD, users can share a network cable, however only one user can transmit data at a specific time.

incoming call restriction - In telephone call-processing feature that disables a telephone from receiving incoming calls.

Incumbent Local Exchange Carrier (ILEC) - A telephone carrier (service provider) that was operating a local telephone system prior to the divestiture of the AT&T bell system.

information highway - A term that refers to a common communication path (highway) for the transport of information. The world wide web (Web) is sometimes called the Information Highway.

Institute of Electrical and Electronics Engineers (IEEE) - An organization formed in 1963 that represents of electrical and electronics scientists and engineers. The IEEE resulted from the merger of the institute of Radio Engineers (IIRE) and the American Institute of Electrical Engineers (AIEE). Its has various societies that focus on key industry technical specialties (such as communications and robotics).

Integrated Access Device (IAD) - A device that converts multiple types of input signals into a common communications format. IADs are commonly used in PBX systems to integrate different types of telephone devices (e.g. analog phone, digital phone and fax) onto a common digital medium (e.g. T1 or E1 line).

integrated digital loop carrier (IDLC) - IDLC systems are the integration of the integrated digital terminal (IDT) and remote digital terminal (RDT). The IDT is part of the local digital switch (LDS) and it acts like a concentrator to put more channels on a digital communications line. The IDLC system moves some of the switching services from the local switches into RDTs to increase the efficiency of communication lines between customers and the central office.

BellCore's (now Telcordia Technologies) GR-303 specification defines the interconnection of the LDS and RDT.

integrated digtal terminal (IDT) - An electronic assembly that is part of a local digital switch (LDS) that coordinates communication with an remote digital terminal (RDT). The IDT concentrates some of the communication channels onto high-speed digital lines that are routed to RDTs. The RDTs demultiplex the digital line and assign channels to individual access lines.

Integrated Services Digital Network (ISDN) - A structured all digital telephone network system that was developed to replace (upgrade) existing analog telephone networks. The ISDN network supports for advanced telecommunications services and defined universal standard interfaces that are used in wireless and wired communications systems.

Intelligent Network (IN) - A telecommunications network architecture that has the ability to process call control and related functions via distributed network transfer points and control centers as opposed to a concentrated in switching system. (*See* Advanced Intelligent Network.)

Inter Symbol Interference (ISI) - Interference to a digital modulated signal that results from the reception of the same radio signal where one of the signals is delayed compared to the other. This is typically caused by multiple receptions of the same signal (multi-path) transmission where a single radio transmitted signal is reflected by objects (such as buildings), and part of the radio energy travels a path of different distance compared to another part of the signal (e.g. direct line of sight compared to reflected off a building or mountain). The interference results when the combined effect of multiple signals changes the decision points used to convert the radio signal back into its original digital form.

interexchange carrier (IXC) - A telephone service company that provides communications services that interconnect local telephone switches. IXCs are commonly called "Long Distance Companies."

interleaving - Interleaving is the reordering of data that is to be transmitted so that consecutive bytes of data are distributed over a larger sequence of data to reduce the effect of burst errors. The use of interleaving greatly increases the ability of error protection codes to correct for burst errors. Many of the error protection coding processes can correct for small numbers of errors, but cannot correct for errors that occur in groups.

International Telecommunication Union (ITU) - A specialized agency of the United Nations established to maintain and extend international cooperation for the maintenance, development, and efficient use of telecommunications. The union does this through standards and recommended regulations, and through technical and telecommunications studies. Based in Geneva, Switzerland, the ITLI is composed of two consultative committees: the International Radio Consultative Committee (CCIIR) and the International Telegraph and Telephone Consultative Committee (CCITT).

Internet - A data network that uses standard Internet protocol (IP) to transfer data throughout the world. The Internet evolved from ARPANET.

Internet engineering task force (IETF) - An organization that assists in the development and coordinates of transaction capability protocol/Internet Protocol standards.

Internet Protocol (IP) - A communications protocol structure that is part of the Transmission Control Protocol and Internet Protocol (TCP/IP). This protocol includes packet delivery addressing, type of service specification, dividing and re-assembly of long data files and data security.

Internet Protocol Security (IPSec) - A part of the Internet Protocol that helps to ensure the privacy of user data. IPSec is part of the next generation internet, IPv6.

Internet service provider (ISP) - A company that provides an end user with data communication service that allows them to connect to the Internet. An ISP purchases a high-speed link to the Internet and divides up the data transmission to allow many more users to connect to the Internet.

Internet Telephony - Telephone systems and services that use the Internet to initiate, process and receive voice communications.

Internet2 - A second generation of the Internet that uses a high-speed backbone communications network. The Internet system is a result of the Next Generation Internet (NGI) initiative that is sponsored by the United States government. Internet2 is seen as the way to deliver multimedia content (e.g. video on demand) through the Internet.

ISDN to POTS converter (IPC) - A device that converts an ISDN basic rate interface (BRI) into a POTS analog telephone interface.

last mile - The last portion of the telephone access line that is installed between a local telephone company switching facility and the customer's premises.

Layer 2 Tunneling Protocol (L2TP) - An protocol that is used for to allow a secure communication path, a virtual private network link, between computers. It is an evolution of earlier point-to-point tunneling protocol (PTPP) as it offers more reliable operation and enhanced security. L2TP enables private communication lines through a public network. L2TP was developed via the Internet engineering task for (IETF).

line coding - The process of modulating and formatting data for transmission on a communications line.

loading coil - A inductive device (temporary storage of energy in a magnetic field) that is installed in a telephone line to help enhance the frequency response of the line at specific audio frequencies. Unfortunately, loading coils significantly add distortion to high-speed data signals on those lines (such as xDSL signals).

Local Access and Transport Area (LATA) - A geographic region in the United States where a local exchange carrier (LEC) is permitted to provide interconnected telephone service. LATAs were created as a result of the division of the company AT&T by the designated by the Modification of Final Judgment (MFJ). A LATA contains one or more local exchange areas, usually with common social, economic, or other interests.

Local Area Network (LAN) - A private network that typically offers high-speed digital communications channels for the interconnection of computers and related equipment in a limited geographic area. LANs can use fiber optic, coaxial, twisted-pair cables or radio transceivers to transmit and receive data signals.

local digital switch (LDS) - A digital switch that is the final switching point between the end customer and the public switched telephone network.

local exchange - Another term for a central office (CO) switching system.

local loop - A connection (wireless or wired) between a customer's telephone or data equipment and a local exchange carrier (LEC) or other telephone service provider. Traditionally, the local loop has been composed of a two or four wire circuit for each telephone line.

logical channel - A portion of a physical communications channel that is used to for a particular (logical) communications purpose. The physical channel may be divided in time, frequency or digital coding to provide for these logical channels.

Main Distribution Frame (MDF) - The wire connection point (wire rack) that is located at or near the central switching that is the point where all local access loops are terminated.

mean time between failures (MTBF) - For a particular time period (typically rated in hours), the total functioning lifetime of an assembly or item divided by the total number of failures for that item within the measurement time interval.

metropolitan area network (MAN) - A data communications network or group of networks that have geographic boundaries of a metropolitan area. The network is totally or partially segregated from other networks, and typically links local area networks (LANs) together.

MODEM - The term is a contraction of MOdulator/DEModulator. It is a device or circuit that converts digital signals to and from analog signals for transmission over conventional analog telephone lines. The term modem also may refer to a device or circuit that converts analog signals from one frequency band to another.

motion picture experts group (MPEG) - An working committee that defines and develops standards for the digital video. These standards define the data compression and decompression processes and how they are delivered on digital broadcast systems. MEG is part of International Standards Organization (ISO) .

MPEG Compression - The compression of video signals as the conform to the motion picture experts group (MPEG). There are various levels of MPEG compression; MPEG-1 and MPEG-2. MPEG-1 compresses by approximately 52 to 1. MPEG-2 compresses up to 200 to 1. MPEG-2 typically provides digital video quality that is similar to VHS tapes with a data rate of approximately 1 Mbps. MPEG-2 compression can be used for HDTV channels, however this requires higher data rates.

multi-party call conference - An enhanced telecommunications service that allows the (or more) users to be connected to the same call.

multicast backbone (MBONE) - A high-speed data communications system that interconnects the Internet that allows multicast services. The MBONE network is composed of interconnected multicast LANs.

multipath - The result of propagation of a radio signals for which part of the signal energy is received before another part of the signal is received that is delayed in time. The delay is due to the extra travel time for the other part of the radio signal that may have been reflected from a building or mountain.

multiple system operator (MSO) - A company that owns more than one telecommunications system that provides communications services. In the United States, MSO is the term that is commonly used to describe a company that owns and operates more than one cable television system.

multiplexer - A device that combines multiple signals onto a single transmission medium. Multiplexing may be in the form of frequency division (e.g. multiple radio channels on a coax line), time division (e.g. slots on a T1 or E1 line), code division (coded channels that share the same frequency band) or combinations of these.

multiplexing - A process that divides a single transmission path to parts that carry multiple communication (voice and/or data) channels. Multiplexing may be time division (dividing into time slots), frequency division (dividing into frequency bands) or code division (dividing into coded data that randomly overlap).

near end cross talk (NEXT) - The leakage of signal that is coupled to a nearby cable or electronics circuit (called crosstalk) where the unwanted signal is received on the originating end (opposite direction) of the cable. NEXT is usually more troublesome than far end crosstalk, as the crosstalk signal levels of NEXT are higher.

near video on demand (NVOD) - A video delivery service that allows a customer to select from a limited number of broadcast video channels when they are broadcast. NVOD channels have pre-designated schedule times and are used for pay-per-view services.

negative acknowledgment message (Nack) - A message that is responded by a communications device to confirm a message or portion of information has NOT been successfully received. If a communications device sends a Nack messages back to the originator, the system will typically re-send the message. See also acknowledgment message (Ack)

network address translation (NAT) - A process that converts network addresses between two different networks. NAT is typically used to convert public network addresses (such as IP addresses) into private local network addresses that are not recognized on the internet. NAT provides added security as computers connected through public networks cannot access local computers with private network addresses.

network interface card (NIC) - The device that adapts computer network protocol to a data bus or data interface in a computer. The NIC between a computer network (such as the Ethernet) and a computer data bus (such as a PCI socket). The NIC is usually a PC expansion board connector and operating system. Software in the computer is installed and setup to recognize the NIC card.

network interface device (NID) - A connection point between the end customers equipment and the telecommunications network. This is also called the demarcation point.

network interface unit (NIU) - An electronic assembly that terminates a telecommunications line to an end user's facility. For optical networks, the NIU may terminate fiber and copper lines and convert the signals into analog (telephone) and digital (computer network or mulit-media) signals.

network maintenance center (NMC) - A facility that allows monitoring, testing and maintenance of a telecommunications network. The NMC is typically operational 24 hours a day, 7 days per week.

network management - A set of procedures, equipment, and operations that keep a telecommunications network operating near maximum efficiency despite unusual loads or equipment failures.

network operator - A company that manages the network equipment parts of a communications system. A network operator does not have to be the service provider. (*See* Service Provider.)

network termination (NT) - A final end point in a network that is usually owned by the network service provider. Wire and telephone equipment that extends from the NT to the end users equipment is commonly owned by the end customer. This is called customer premises equipment (CPE). When the network termination (NT) is an active device, it typically has standard communications parameters such as protocols, timing and voltages to allow specific types of equipment to correctly communicate with the network.

noise - Any random signal or unwanted disturbance in a communication system that tends to interfere with the processing of the desired signal's (non-noise) function.

North American Numbering Plan (NANP) - A telephone numbering system used in North America that uses 10-digit numbering. The number consists of a 3-diigit area code, a 3-digit central office code, and a 4-digit line number.

NTSC signals - The NTSC television system standard was developed in the United States and is used in many parts of the world. The NTSC system uses analog modulation where a sync burst precedes the video information. The NTSC system uses 525 lines of resolution (42 are blanking lines) and has a pixel resolution of approximately 148k to 150k pixels.

off peak - A time period where a telecommunication system usage is lower, typically after normal business hours. Some telecommunications service providers charge a reduced rate for the use of services during off-peak hours.

on peak - A time period where a telecommunication system usage is higher, typically during normal business hours. Some telecommunications service providers charge a premium rate for the use of services during peak hours.

open system interconnection (OSI) - A layered functional model that defines the parts of a telecommunications system so parts can be developed independently by different companies to industry standards.

operations administration and maintenance (OA&M) - The functions that are necessary to operate, perform administration functions and maintain a communications network.

Operations and Maintenance Center (OMC) - The OMC includes alarms and monitoring equipment to help a network operator diagnose and repair a communications network.

operations subsystem (OSS) - The system that is used to allow a network operator to perform the administrative portions of the business. These functions include customer care, inventory management and billing.

optical network unit (ONU) - ONU's are used to multiplex and demultiplex signals to and from a fiber transmission line. An ONU terminates an optical fiber line and converts the signal to a format suitable for distribution to a customer's equipment. When used for residential use, a single ONU can server 128 to 500 dwellings.

Over-subscription - A situation that occurs when a service provider sells more capacity to end customers than a communications network can provide at a specific time period. This provides a benefit of reduced network equipment and operational (reduced leased line) cost.

Over-subscription is a common practice in communications networks as customers do not continuously use the maximum capacity assigned to them and customers access the networks at different time periods. Unfortunately, over-subscription in telecommunications can cause problems when customers do attempt to access the network at the same time. For example, when customers open their presents at a holiday event (e.g. Christmas) and attempt to access the Internet at the same time.

packet - A small group of digital bits that is routed through a network to their destination. The bit sequence of the packet (field structure) of a packet may be arranged to include the destination address in addition to the data that is being transported and other data such as the packet originator and error protection bits.

packet assembler and disassembler (PAD) - A device that divides or converts blocks of data (such as data files) to and from small packets of information. In the disassembly process, a PAD usually assigns sequential numbers to the packets as they are created to allow the re-assembly PAD to identify the correct sequence to reproduce the original data signal.

packet assembler/disassembler (PAD) - A software process that breaks data messages into segments for transmission in short bursts through a packet switching network and reassembles the packets at the receiving end to create the original data message. When the PAD disassembles the message, it adds codes for address, routing, and error detection.

packet switched data - Packet switch data is the transfer of information between two points through the division of the data into small packets. The packets are routed (switched) through the network and reconnected at the other end to recreate the original data. Each data packet contains the address of its destination. This allows each packet to take a different route through the network to reach its destination.

Packet Switching - A mode of data transmission in which messages are broken into increments, or packets, each of which can be routed separately from a source then reassembled in the proper order at the destination.

paging - A method of delivering a message, via a public communications system or radio signal, to a person whose exact whereabouts are unknown by the sender of the message. Users typically carry a small paging receiver that displays a numeric or alphanumeric message displayed on an electronic readout or it could be sent and received as voice message or other data.

passband - A range of frequencies that can be passed through a filter or medium essentially unchanged.

pay-per-view - A video delivery service that allows customers to select and pay for individual movie or video selections. Pay-per-view services can allow access to movies at any time (video on demand) or may only offer access at pre-scheduled intervals (near video on demand).

per call block - A feature of calling number identification that allows the originator of a call to disable the display of their calling number. This feature may be used when a caller wishes to place an anonymous complaint.

per line block - A feature of calling number identification that allows the user of the telephone device to disable the display of their calling number when making calls. This feature may be used when a caller wishes to place calls from a private number.

peripheral component Interconnect (PCI) - A standard data communication connection that allows accessory cards to be installed into a personal computer. The PCI specification defines both the electrical and physical (connector) requirements. PCI was introduced by Intel and it allows up to 10 PCI-complient expansion cards in a PC. The PCI standard has replaced the previous industry standard architecture (ISA) bus.

Personal Firewall - A device or software program that runs on your computer that provides protection from Internet or data network intruders. Firewalls can restrict access types and may monitor for advanced security threats by analyzing certain types of data communication activities. Although firewalls are important for protecting data that is connected to pubic networks, they can be complicated to setup, can cause problems with desired communications and generally can slow down the transfer of data communication.

personal identification number (PIN) - A number assigned to an individual subscriber which is used to gain access to specified services, such as credit card calling or prepaid wireless services.

phase modulation (PM) - A modulation process where the phase of the radio carrier signal is modified by the amplitude of the information (typically audio) signal.

phone doubler(TM) - The telephone service that provides a customer an indication that an incoming call (usually a voice call) is waiting for them while they are connected to the Internet. The service allows the customer to temporarily hold the Internet connection while they communicate on the other communications channel (answer the voice call). When the customer hangs up the phone, the Internet connection is restored and Internet service resumes.

Phoneline network - A computer network technology that allows standard telephone wiring to be used as network cabling without the need to disconnect standard telephones. The Phoneline Network uses high frequency signals that are above standard telephone and DSL frequency bands. To install a phone line network, end users install Phoneline NICs in a similar method to adding an EtherNet card. The Phoneline networking system allows computers to be connected to each other without the use of a hub (daisy chain).

plain old telephone service (POTS) - Basic telephone service without any enhanced features. It is the common term for residential telephone service.

Plug-and-Play (PnP) - A compatibility system that simplifies the installation and removal of hardware devices into a personal computer. PnP includes the automatic recognition of hardware installation and removal, activation and deactivation of software drivers, and system accessory management functions.

Point-to-Point Protocol (PPP) - A communications protocol that uses Transmission Control Protocol and Internet Protocol (TCP/IP) to allow end users (end points) to connect directly to the Internet via a communications connection (usually dialup connection). For Microsoft operating systems, this type of connection is managed by Dial-up networking (DUN).

Point-to-point Tunneling protocol PPTP - Point-to-point tunneling protocol (PPTP) is a virtual private network solution. PPTP software is included with Microsoft windows95, 98, NT, and 2000 and is built into some routers.

PPTP is an easy protocol to setup. However its first revision does not provide for data encryption. This lowers the privacy for data transmission.

Powerline Ethernet - A transmission protocol that uses allows data transmission on standard electric (typically residential) power lines. Powerline Ethernet is a packet switched technology that is used for local area networks (LANs). Ethernet is a registered trademark of Xerox Corporation.

premises distribution network (PDN) - The network used at the customer's facility to connect terminals to other networks and each other.

Primary Rate Interface (PRI) - An standard high-speed data communications interface that is used in the ISDN system. This interface provides a standard data rates for T1 1.544 Mbps and E1 2.048 Mbps. The interface can be divided into combinations of 384 kbps (H) channels, 64 kbps (B) channels and includes at least one 64 kbps (D) control channel.

private branch exchange (PBX) - A private telephone system that is used to provide telephone service within a building or group of buildings in a small geographic area. PBX systems contain small switches and advanced call processing features (such as speed dialing, call transfer) and are typically connected to the public switched telephone network (PSTN).

Protocol - (1 - rules) A precise set of rules and a syntax that govern the accurate transfer of information. (2- connection) A procedure for connecting to a communications system to establish, carry out, and terminate communications.

provisioning - The process that a communications service performs to add, update and remove customers to their authorized list of service users. Provisioning includes processing new service orders, installation and truck dispatch orders, customer service functions for adjusting billing charges. Provisioning is also called customer care.

PSTN gateway - A PSTN gateway is a communications device or that transforms data that is received from one network (such as the Internet or DSL network) into a format that can be used by the PSTN network. The PSTN gateway may be a simple device that performs simple call completion and adaptation of digital audio into a compatible signals for the PSTN. Or, it may be a more complex device that is capable of advanced services such as conference calling, call waiting, call forward and other PSTN like services. The PSTN gateway must create signaling protocols and compensate for timing differences between a end users computer and the public switched telephone network (PSTN).

Public Switched Telephone Network (PSTN) - An unrestricted dialing telephone network that is available for public use. The network is an integrated system of transmission and switching facilities, signaling processors, and associated operations support systems that is shared by customers. PSTN is also called a public network, public switched network, or public telephone network.

public telephone network (PTN) - An unrestricted dialing telephone network that is available for public use. The network is an integrated system of transmission and switching facilities, signaling processors, and associated operations support systems that is shared by customers. PTN is the common term used for public telephone networks outside North America. Inside North America, the public telephone network is called the public switched telephone network. In the United States, the PTN is referred to as the Public Switched Telephone Network (PSTN).

Quadrature Amplitude Modulation (QAM) - A process that enables two signals to modulate a single carrier frequency. The two different signals' amplitude modulates two samples of the carrier that are of the same frequency, but differ in phase by 90 degrees. The resultant two signals can be added together and both signals recovered at a decoder where they are then demodulated 90 degrees apart.

quality of service (QoS) - One or more measurements of desired performance of a communications system. QoS measures may include availability, maximum bit error rate (BER), minimum committed bit rate (CBR) and other measurements that are used to ensure quality telecommunications service.

Rate Adaptive Digital Subscriber Line (RADSL) - A hybrid analog and digital subscriber line technology that allows the data transmission rate to dynamically change dependent on the electrical transmission characteristics and/or the settings provided by the DSL service provider. RADSL features have been incorporated into standard ADSL technology.

rate plan - The structure of service fees that a user will pay to use telecommunications services. Rate plans are typically divided into monthly fees and usage fees.

redundancy - A system design that includes additional equipment for the backup of key systems or components in the event of an equipment or system failure. While redundancy improves the overall reliability of a system, it also increases the number of equipment assemblies contained within a network. Redundancy usually increases cost.

regional bell operating company (RBOC) - A United States telephone company that is one of the seven telephone companies that were created as a result from the division of AT&T in 1983. RBOCs are also known as the Baby Bells.

registration - (1- FCC) A legally required procedure whereby vendors must submit their telephone equipment for testing and certification before it can be directly connected to a public telephone network. (2- wireless system notification) A process where a mobile radio transmits information to a wireless system that informs it that it is available and operating in the system. This allows the system to send paging alerts and command messages to the mobile radio.

regulation - (1- electrical quantity) A process of adjusting the parameters of some signal or system (such as circuit gain). (2 - government) Rules established by a government agency that are designed to maintain the service public communications systems.

reliability - The ability of a network or equipment to perform within its normal operating parameters to provide a specific quality level of service. Reliability can be measured as a minimum performance rating over a specified interval of time. These parameters include bit error rate, minimum data transfer capacity or mean time between equipment failures (MTBF).

remote digital terminal (RDT) - The RDT provides an interface between a high-speed digital transmission line (e.g. DS1) and the customer's access line. The RDT can dynamically assign time slots from a high-speed line to customer access lines. Because customer access lines are not used at the same time, an RDT that interfaces to a DS1 line usually provides service to 96 customer access lines.

The RDT is divided into three major parts; digital transmission facility interface, common system interface and line interface. The digital transmission interface terminates the high-speed line and coordinates the signaling. The common system interface performs the multiplexing/de-multiplexing, signaling insertion and extraction. The line interface contains digital to analog conversions (if the access line is analog) or digital formatting (if the line is digital).

Research in Advanced Communications in Europe (RACE) - A cooperative research program started in Europe commissioned to develop the technology for Broadband Integrated Services Digital Network (B-ISDN) systems. The RACE members are reviewing a Mobile Broadband System operating in the 60 GHz bands for mobile service applications in the approximate range of 2-100 Mb/s.

retrieval services - A service that allows authorized users of the service to retrieve information from an information center.

ring tone - The tone that is used to announce an incoming telecommunications call. There may several different types of ring tones and some telecommunications devices allows the user to program in their own unique selection for a ring tone.

RJ-45 - A standard 8 wire modular connector. RJ-45 connectors are commonly used in telephone and data communication systems.

router - A router is a device that directs (routes) data from one path to another in a network. Routers base their switching information on one or more information parameters of the data messages. These parameters may include availability of a transmission path or communications channel, destination address contained within a packet, maximum allowable amount of transmission delay a packet can accept along with other key parameters. Routers that connect data paths between different types of networks are sometimes called gateways.

sampling - The process of taking samples of an electronic signal at equal time intervals to typically convert the level into digital information.

scrambling - A process of altering or changing an electrical signal (often the distortion of a video signal) to prevent interpretation of the signals by users that can receive the signal but are unauthorized to receive the signal. Scrambling involves the changing of a signal according to a known process so that the received signal can reverse the process to decode the signal back into its original (or close to original) form.

sealing current - A method that is used to decrease the effects of corrosion is to continuously run current through the copper wire pair. This "sealing current" is a small amount of direct current that is passed through a copper wire to reduce the corrosion effects of the splice points. The sealing current effectively maintains conductivity of mechanical splices that are not soldered. The direct current effectively punches holes the corrosive oxide film that forms on the mechanical splices.

server - A computer that can receive. process and respond to an end user's (client's) request for information or information processing.

Service Provider - An generic name given to a company or organization that provides telecommunications service to customers (subscribers). (*See* Network Operator.)

set top box - An electronic device that adapts a communications medium to a format that is accessible by the end user. Set top boxes are commonly located in a customers home to allow the reception of video signals on a television or computer.

shielded twisted pair (STP) - Wire that includes twisted pairs and a shield that surrounds the twisted pairs of wires. STP uses a metal shield (foil) that surrounds the twisted pair wires to help protect against unwanted electromagnetic interference.

Short Message Service (SMS) - A messaging service that typically transfers small amounts of text (several hundred characters). Short messaging services can be broadcast without acknowledgement (e.g. traffic reports) or sent point-to-point (paging or email). Most digital cellular systems have SMS services. Short messaging for mobile telephones may include: numeric pages (dialed in by a caller), messages that are entered by a live operator via keyboard, an automatic message service that sends a predefined message when an event occurs (such as a fire alarm or system equipment failure), network operator announcements to customers, to and from other message capable devices in the system, from the Internet, advertisers or other information providers.

signal egress - The emission of a portion of a transmitted signal from a transmission line. The emission is usually unwanted and may cause interference to neighboring cables or electronic circuits.

signal ingress - The absorption of radio signal energy from an external source into a communications circuit or communications link. Signal ingress may occur when electrical signals from sources such as radio or lightning spikes occur.

signaling - The process of transferring control information such as address, call supervision, or other connection information between communication equipment and other equipment or systems.

signaling point (SP) - In the Signaling System 7 protocol, a node in a signaling network that originates and receives signaling messages, transfers signaling messages from one signaling link to another, or both.

Signaling System #7 (SS7) - A international standard network signaling protocol which allows common channel (independent) signaling between telephone network elements. Because the SS7 system protocols have been optimized for telephone system control connections, the SS7 system can typically efficiently process more control messages than other packet systems such as the X.25 system.

Signaling Transfer Point (STP) - A signaling switch used in the SS7 common channel signaling network. These transfer points are used to route signaling messages (packets) to other signaling transfer points or network parts.

Simple Network Management Protocol (SNMP) - A communication protocol that is used to manage network equipment. By conforming to this protocol, equipment assemblies that are produced by different manufacturers can be managed by a single program. SNMP protocol can operate via Internet protocol. The communication principles of SNMP are similar to those provide by IBM's NetView.

soft keys - Soft keys are buttons on the keypad of an electronic device that have the ability to redefine their functions. Soft keys are typically located adjacent to a display that provides a description of the key function. This allows an electronic assembly to reduce the number of keys which is especially important for portable handheld telephones.

speech coder - A data compression device that characterizes and compresses digital speech information. (*See* CODEC.)

splitter - A circuit, device or component that divides a complex input signal into several outputs. A splitter may simply divide the signal energy power from a single source to 2 or more outputs or the splitter may separate frequency components to different output ports.

spread spectrum - A method of spreading information signals (typically digital signals) so the frequency bandwidth of the radio channel is much larger than the original information bandwidth.

Static IP Addressing - An process of assigning a fixed Internet Protocol (IP) address to a computer or data network device. Use of a static IP address allows other computers to initiate data transmission (such as a video conference call) to a specific recipient.

subnet mask - The assignment of a number that is used to separate sections of a network by using a specific sequence of numbers in a network address. When used with the internet, a subnet mask is 32 bits.

subscriber - The end user of telecommunications services.

subscriber identity module (SIM) - The subscriber identity module (SIM) is a small "information" card that contains service subscription identity and personal information. This information includes a phone number, billing identification information and a small amount of user specific data (such as feature preferences and short messages). This information is stored in the card rather than programming this information into the phone itself. This intelligent card, either credit card-sized (ISO format), or the size of a postage-stamp (Plug-In format), can be inserted into any SIM ready wireless telephone.

subscriber interface unit (SIU) - An electronic assembly that is used to convert a digital signal (such as from a network interface unit) to a suitable format for a customers user (such as a television signal).

switch - A network device (typically a computer) that is capable of connecting communication paths to other communication paths. Early switches used mechanical levers (cross-bars) to interconnect lines. Most switches use a time slot interchange memory matrix to dynamically connect different communications paths through software control.

Symmeretical Digital Subscriber Line (SDSL) - An all digital transmission technology that is used on a single pair of copper wires that can deliver near T1 or E1 data transmission speeds. SDSL is a symmetrical service that ranges from 160 kbps to 2.3 Mbps and can reach to 18000 feet from the central switching office.

synchronous optical network (SONET) - A digital transmission format that is used in optical (fiber) networks to transport high-speed data signals. SONET uses standard data transfer rates and defined frame structures formats in a synchronous (sequential) format.

System Network Architecture (SNA) - A communications system protocol developed by IBM that allows for control and data communication via different types of network communication equipment.

T1 carrier - 24 voice channels digitized at 64,000 bps, combined into a single 1.544 Mbps digital stream and carried over two pairs of regular copper telephone wires. Now used for dedicated local access to long distance facilities, long haul private lines, for regular local service or for resellers interconnecting to carrier switches.

Telecommunications Act of 1996 - The U.S. Telecommunications Act of 1996 provides a national framework for the deregulation of the local exchange market, a deregulation that was already taking place on a state-by-state basis through the actions of state regulatory commissions. Its summary impact on the local exchange market is to require current LECs to remove all barriers to the competition (e.g. interconnect, white and/or yellow pages access, co-location, & wholesaling of facilities restrictions) in return for LEC access to the long distance market.

telecommuting - The process of an employee that is conducting business related activities at a remote location (usually at a home) through the use of telecommunications services and equipment. Telecommuting allows employees to work at home without the need to commute to the office and reduces the need for the business to maintain office space for workers.

teleconferencing - A process of conducting a meeting between two or more people through the use of telecommunications circuits and equipment. Teleconferencing usually involves sharing video and/or audio communications.

telemedicine - Processes that assist with health care service that employ communications services and equipment. Examples of telemedicine include delivery of medical images, remote access to medical records, remote monitoring of heath care equipment equipment and distant monitoring of biological functions such as heart rate and blood pressure.

telemetry - The transfer of measurement information to a monitoring system through the use of wire, optical fiber, or radio transmission. The term telemetry is often used with the gathering of information.

telephony application programming interface (TAPI) - An industry standard that allows computers to access and control telecommunications devices.

teleservices - Telecommunications services process or store user data as it is transported through a communications network. An example of a teleservice is a fax forward and storage service.

time division duplex (TDD) - Time division duplex (TDD) is a process of allowing two way communications between two devices by time sharing. When using TDD, one device transmits (device 1), the other device listens (device 2) for a short period of time. After the transmission is complete, the devices reverse their role so device 1 becomes a receiver and device 2 becomes a transmitter. The process continually repeats itself so data appears to flow in both directions simultaneously.

Time Division Multiple Access (TDMA) - Time division multiple access (TDMA) is a process of sharing a single radio channel by dividing the channel into time slots that are shared between simultaneous users of the radio channel. When a mobile radio communicates with a TDMA system, it is assigned a specific time position on the radio channel. By allow several users to use different time positions (time slots) on a single radio channel, TDMA systems increase their ability to serve multiple users with a limited number of radio channels.

time slot - The smallest division of a communications channel that are assigned to particular users in the system. Time slots can be combined for a single user to increase the total data transfer rate available to that user. In some systems, time slots are assigned dynamically on an as-needed basis.

Time Slot Interchange (TSI) Switching - A process of connecting incoming and outgoing digital lines together through the use of temporary memory locations. A computer controls the assignment of these temporary locations so that a portion of an incoming line can be stored in temporary memory and retrieved for insertion to an outgoing line.

token - The code passed among nodes (typically computers) in network in a particular sequence for each node. This sequence indicates which note has permission to transmit data next.

training sequence - A sequence of data bits that are previously known to the sender and receiver of the data bits. This allows the receiver to adjusts its reception process by using the known sequence.

Transmission Control Protocol and Internet Protocol (TCP/IP) - TCP/IP is standard set (suite) of protocols that defines how the internet messages are transferred reliably. The Transmission Control Protocol (TCP) portion ensures message delivery between two points and the Internet Protocol (IP) defines the routing and physical of packets of data.

transmit delay - The time delay between when a signal is first originated to when it is first received at its destination. Also called transmission time or propagation delay for radio signals.

trunk - A communication path that connects two network elements (such as switching systems, networks or data devices). Trunks are shared by many users and they are often classified by the type of equipment they connect. For example, a PBX trunk connects a PBX system to the public switched telephone network (PSTN).

Trunks - Groups of wires or fiber optic communication lines that are used to interconnect communications devices.

unbundling services - The process of separating portions of a telecommunications services that are provided by a network. Unbundling is a common term used to describe the separation of standard telephone equipment and services to allow competing telephone service providers to gain fair access to parts of incumbent telephone company systems. An example of an unbundled service is for the incumbent phone company to lease access to the copper wire line that connects an end user to the local telephone company. The competing company may install high-speed data modems (such as ADSL) on the copper line to enhancing the value of the telecommunications service.

Universal ADSL Working Group (UWAG) - A working group that was established to assist in the standardization of ADSL equipment for the consumer marketplace. The UWAG was setup primarily by telecommunications service providers and computer equipment manufacturers.

Universal Resource Locator (URL) - A standardized addressing process used to the Internet.

universal serial bus (USB) - An industry standard data communication interface that is installed on personal computers. The USB was designed to replace the older UART data communications port. USB ports permit data transmission speeds up to 12 Mbps and up to 10 devices can share a single USB port.

unshielded twisted pair (UTP) - The transmission line for xDSL systems is typically unshielded twisted pair (UTP). UTP consists of pairs of copper wires twisted around each other and covered by plastic insulation. The twisting of the copper wire pair reduces the effects of interference as each wire receives approximately the same level of interference (balanced) thereby effectively canceling the interference. UTP is by far the most popular cabling used for local access lines and computer LANs (such as 10BaseT and 100BaseT).

unspecified bit rate (UBR) - A category of telecommunications service that provide an unspecified data transmission rate of service to end user applications. Applications that use UBR services do not require real-time interactivity nor do they require a minimum data transfer rate. An example of a UBR application is Internet web browsing.

uplink - (1- Satellite) The earth-to-satellite microwave link and related components such as earth station transmitting equipment. The satellite contains an uplink receiver. Various uplink components in the earth station are involved with the processing and transmission of the signal to the satellite. (2- cellular systems) The radio link between the mobile station and the base station.

user - A person, company, or group that uses the services of a system for the transfer of voice, data information or other purposes.

user network interface (UNI) - The interface between an end user and a telecommunications network. A UNI could be a industry standard set of protocol rules and data transmission specifications or may be a proprietary protocol.

variable bit rate (VBR) - A category of telecommunications service that provide an variable data transmission rate of service to end user applications. Applications that use VBR services usually require some real-time interactivity with bursts of data transmission. An example of a VBR application is videoconferencing.

very high bit rate digital subscriber line (VDSL) - A communication system that transfers both analog and digital information on a copper wire pair. The analog information can be a standard POTS or ISDN signal and the typical downstream digital transmission rate (data rate to the end user) can vary from 13 Mbps to 52 Mbps downstream and the maximum upstream digital transmission rate (from the customer to the network) can be 26 Mbps. The data transmission rate varies depending on distance, line distortion and settings from the VDSL service provider. The maximum practical distance limitation for VDSL transmission is approximately 4,500 feet (~1,500 meters). However, to achieve 52 Mbps, the maximum transmission length is approximately 1,000 feet (~300 meters).

Video Mail (VMail) - A process of recording and sending short video messages (typically 1-2 minute video clips) in digital form via an electronic mail (email). Video mail messages may be sent as an attachment to standard Email addresses.

video on demand (VOD) - A service that provides end users to interactively request and receive video services. These video services be from previously stored media (entertainment movies or education videos) or have a live connection (news events in real time).

video telephony - A telephone service that allows customers to hear and see another telephone user or video source. Video telephony applications include video on demand (VOD) movies, distance learning (remote education), telemedicine, teleconferencing and other applications that can benefit from the combination of video and audio signals.

virtual circuit or virtual channel (VC) - A logical connection between two communication ports in one or more communication network. Virtual circuits are used to temporarily connect data terminals to host computers. Because virtual circuits logically connect communication ports together, a single network switching system may be used to provide for many virtual circuits.

Virtual Private Network (VPN) - Secure private communication path(s) through one or more data network that is dedicated between two points. VPN connections allow data to safely and privately pass over public networks (such as the Internet). The data traveling between two points is encrypted for privacy.

voice activity detector (VAD) - An electronic circuit that senses the activity (or absence) of voice signals. This is often used to inhibit a transmission signal during periods of voice inactivity.

voice coder (Vo-coder) - A digital signal processing device that analyzes voice signals so that it can produce a lower data rate compressed digital signal.

Voice Mail (VM) - A service that provides a telephone customer with an electronic storage mailbox that can answer and store incoming voice messages. Voice mail systems use interactive voice response (IVR) technology to prompt callers and customers through the options available from voice mailbox systems. Voice mail systems offer advanced features not available from standard answering machines including message forwarding to other mailboxes, time of day recording and routing, special announcements and other features.

voice mail system (VMS) - A telecommunications system that allows a subscriber to receive and play back messages from a remote location (such as a PBX telephone or mobile phone).

voice over DSL (VoDSL) - Sending voice over a digital subscriber line system (VoDSL) is a process that sends audio band (also called "voice band") signals (e.g. voice, fax or voice band modem) via a digital channel on a digital subscriber line (xDSL) system. VoDSL requires conversion from analog signals to a digital format and involves the formatting of digital audio signals into frames and time slots so they can be combined onto a digital (DSL) channel.

To communicate to other users, VoDSL requires one or more communication device that are capable of sending and receiving with the DSL network and conversion of a digital channel back into its analog voice band signal. This can be as simple as a computer with a sound card, a DSL modem and VoDSL software or as complex as a companies telephone network with an integrated access device (IAD). Optionally, some DSL systems have a PSTN gateway that can convert digital audio on a DSL system into telephone signals that can be sent through the public switched telephone network.

Voice over internet protocol (VoIP) - A process of sending voice telephone signals over the Internet. If the telephone signal is in analog form (voice or fax, the signal is first converted to a digital form. Packet routing information is then added to the digital voice signal so it can be routed through the Internet.

voice recognition - A computer-based technology that analyzes audio signals (typically spoken words) converts them into digital signals for other processing (e.g. voice dialing).

Wavelength Division Multiplexing (WDM) - A process of using transmitting several high speed communication channels through a single fiber through the use of separate wavelengths (optical frequencies) for each communication channel.

WebCAM - A digital camera that provides images to the Internet. Some WebCams provide continuous image updates while others refresh their images every few seconds to reduce the data transmission requirement.

whiteboard - A device that can capture images or hand drawn text so they can be transferred to a video conferencing system. Whiteboards allow video conferencing users to place share documents, images and/or hand written diagrams with one (or more) video conference attendees.

wide area network (WAN) - A communications network serving geographically separate areas. A WAN can be established by linking together two or more metropolitan area networks, which enables data terminals in one city to access data resources in another city or country.

wireless cable - "Wireless Cable" is a term given to land based (terrestrial) wireless distribution systems that utilizes microwave frequencies to deliver video, data and/or voice signals to end-users. There are two basic types of wireless cable systems, Multichannel Multipoint distribution Service (MMDS) and Local Multichannel Distribution Service (LMDS).

world wide web (WWW) - A service that resides on computers that are connected to the Internet that allows end users to access data that is stored on the computers using standard interface software (browsers). The WWW (commonly called the "web") is associated with customers that use web browsers (graphic display software) to public users to find, acquire and transfer information.

X.25 - An international standard for communications with a packet data switching network. The X,25 standard specifies the protocol between the data device (such as a computer) and the network (such as a public packet data network).

xDSL - A set of large-scale high bandwidth data technologies that can use standard twisted-pair copper wire to deliver high-speed digital services (up to 52 Mbps).

xDSL Splitter - A circuit, device or component that divides an xDSL signal into separate voice and data outputs. An xDSL splitter is typically used for ADSL and VDSL systems.

The xDSL splitter separates the existing telephone signal from the high-speed data signal. In the United States, the standard telephone signal (POTS) frequency band extends up to 8 kHz. In Europe, standard telephone signals include additional high frequency components that extend up to 12 kHz. When the xDSL splitter is used to allow ISDN signals, the frequency band for the ISDN signal extends up to 80 kHz (120 kHz for ISDN in Germany).

Appendix III

United States xDSL Providers

DSL carriers provide have expertise and information that may help you decide on the DSL solution that is best suited for your applications. Below is a list of DSL carriers in the United States.

Ameritech
web: www.ameritech.net
800-910-4369

Bell Atlantic
web: www.bell-atl.com/infospeed, www.bell-atl.com/smallbiz
877-525-2375

Bell South
web: consumer.bellsouth.net/adsl, dsl.smlbiz.bellsouth.com

Bluestar Communications
web: www.bluestar.com

Choice One Communications
web: www.choiceonecom.com
716-246-4231, 888-832-5800

Cincinnati Bell
web: www.cinbelltel.com
513-566-9666

Covad Communications
web: www.covad.com
800-462-6823

Florida Digital Network
web: www.floridadigital.net
877-225-5336, 877-433-6435

Vorizon - GTE
web: www.gte.net
800-483-4000

HarvardNet
web: www.harvardnet.com
800-772-6771, 617-242-1700

InterAcess
web: www.interaccess.com
312-496-4200

Jato Communications
web: www.jato.net
877-452-8663
303-483-7500

Network Access Solutions
web: www.nas-corp.com
888-633-3375

Nevada Bell
web: public.nvbell.net
775-323-2375

New Edge Networks
web: www.newedgenetworks.com
360-693-9009

NorthPoint Communications
web: www.northpoint.net
800-981-8213, 877-836-7375

Pacific Bell
web: www.pacbell.com
888-884-2375

Rhythms NetConnections
800-749-8467

SNET
web: www.snet.com/adsl
877-999-9375

Southwestern Bell
web: www.swbell.com/dsl
888-792-3751

Sprint
web: www.sprint.com
877-746-8466

US West
web: www.uswest.com/products/data/dsl
877-665-6342

Vitts
web: www.vitts.com
888-656-1800

Index

100BaseT, 109, 111

Access lines, 20-21, 35, 37, 49, 107, 111, 113, 130, 139, 143, 163, 191

ADSL
downloading, 190, 192
downstream, 25, 46-47, 49-51, 159, 168
equipment, 2, 19-20, 22-23, 25, 39, 41-42, 44, 48-50, 59-61, 65, 70, 72, 75-76, 78-80, 83-86, 101-103, 107-108, 110-111, 114, 116, 118-139, 141, 143, 145-148, 150, 163, 167-168, 177, 187, 191-192, 196, 199, 206, 210
forum, 47-48, 51, 53
G.lite, 49, 103
Lite, 49, 103
modulation, 8, 17-18, 24-25, 41, 45, 48, 53, 83, 85, 153, 156
upstream, 25, 46-47, 49-51, 168

Advanced Intelligent Network (AIN), 11, 7

Always-on connection

security, 15, 10-11, 62, 95-97, 99-101, 173, 175, 179, 182, 193

American wire gauge (AWG), 103

asynchronous transfer mode (ATM), 30, 47, 51, 83, 90-92, 107, 112, 114-116, 128, 136, 138, 140, 146, 163, 175-177, 193

ATM
cell, 72-73
forum, 47-48, 51, 53

attenuation, 18, 31-32, 36-38, 157-159

availability, 1, 18, 53, 114, 116, 119, 128, 130, 146-147, 149-150, 182, 191

backbone, 59, 69, 107, 112-113, 116, 120, 136, 147, 163, 176

backbone networks, 116, 176

bandwidth
considerations, 146-147
multiple connections, 18
needs, 7, 13, 5, 9, 91, 98, 119, 180, 184

servers, 95

video conferencing, 146, 168, 170-172, 184, 203, 209-210

bandpass, 103

basic rate interface (BRI), 19

bearer services, 163

bit rate, 43, 45, 52, 176, 178

bits, 24, 27-29, 156, 173

bridge, 12, 35-36, 45-46, 85, 89, 92, 107, 109, 119, 161, 172

bridge taps, 35-36, 46, 119

bridged service

firewalls, 99-100, 170, 175

Broadband

Business, 9, 11, 6, 8, 49, 51, 54-55, 59, 67, 80, 103, 109, 115, 127, 133, 135, 143-144, 159-160, 180-181, 183, 186-187, 189, 193, 197, 203-205

Downloading, 190, 192

e-mail, 101, 172, 179-180, 185, 203, 205, 212-213

Online banking, 193-194

Speed, 2, 12, 14, 16, 2, 9-10, 12, 19, 21-23, 26-27, 40, 42-46, 50-51, 53-55, 57, 67, 69, 74-75, 77, 79, 104, 120, 131, 134-135, 144-145, 153, 161, 163, 169, 175-177, 179-181, 183-184, 189, 196, 204, 210, 213

Telecommuting, 203

Video, 1-2, 13, 15-16, 1-2, 9, 11, 27, 29, 53, 56-57, 59, 61-67, 74, 78, 80-81, 89, 91-92, 115-116, 140, 146-147, 163, 168-172, 175-176, 178-184, 187, 190-191, 195-199, 201, 203, 206-210, 212, 214

Video conferencing, 146, 168, 170-172, 184, 203, 209-210

broadcast

fax, 67, 95, 165-166, 212

messaging, 19, 75, 78, 197, 212-213

service, 2, 11, 13-14, 16, 2, 8, 10-12, 19, 22, 29, 35, 42, 46-49, 56-67, 73-76, 79-81, 84-85, 92, 95, 101-104, 107-108, 110-111, 115-122, 127, 133-134, 136-149, 158-159, 163-166,

168, 170, 173-177, 180-182, 199, 201, 203, 205-207, 211-212

business

broadband, 55, 161, 179-184, 187, 189-190, 192, 196, 198-199, 202-203, 206, 211, 213-214

research, 121-122, 188, 193, 195, 199, 210, 213-214

routers, 85, 107, 114, 116, 128, 130, 170, 173, 176

telecommuting, 203

bytes, 142

cable access television (CATV), 55, 57-60, 118, 140

cable modem, 56-57

cabling

connectors, 22, 104-105, 123, 131

installation, 4, 12, 3, 9, 18, 20, 22, 32, 42, 44, 49, 54, 79-80, 87, 96, 102-104, 125, 136-137, 139, 150, 158, 161, 167

carrier, 9, 6, 17-18, 20-21, 23-25, 47-48, 80-81, 120, 147, 150, 156

carrierless amplitude and phase modulation (CAP), 24-25

central office (CO), 3-4, 9, 12, 3, 6, 8-9, 20, 35-36, 41, 45, 47, 49, 53, 61, 67, 80, 108, 112, 114, 128, 135-138

channel, 10-11, 7, 17, 19-21, 27, 29-30, 33, 43, 46, 53, 56-57, 61, 63, 69, 71, 78, 80, 83, 85, 89, 92, 95, 112, 114, 136-137, 140, 159, 165, 167-169, 173, 178, 193, 197, 211

channel capacity, 29, 61, 63

circuit switched data, 145

circuits, 4, 10, 2-3, 7, 22, 44, 53, 79-80, 96, 123, 130, 154, 157, 211, 213

Classes, 64, 76, 183, 185

commerce, 179, 182, 192

communications

asymmetric, 46

digital, 2, 5-7, 13-16, 2, 5, 9-12, 17-24, 27-29, 31, 33, 39-50, 52-54, 56, 60-61, 64, 70, 74-75, 78, 80, 83-84, 89, 91-92, 95, 103-105, 108-

109, 111-113, 116, 120, 123, 128, 130, 136, 140, 143, 146, 150, 153-156, 159, 161, 165-170, 172-173, 176, 178, 180-182, 190-191, 199-200, 202-203, 205-207, 211

symmetric, 45

competition

cable modems, 15, 11, 56-57, 127, 146

standards, 39, 74, 84-85, 122, 128, 146, 169-170, 180

competitive access provider (CAP), 24-25, 48

competitive local exchange carrier (CLEC)

competition, 15, 11, 59, 65, 118, 120, 127, 146-147, 158

data, 1-2, 8, 11-16, 1-2, 6-7, 9-12, 18-19, 21-27, 29-30, 34, 39, 41-59, 61, 64-75, 78-79, 83-88, 90-92, 94-96, 98-100, 104-105, 109, 112-121, 128, 134-140, 144-148, 150, 153-156, 158-163, 165, 167-170, 172-176, 178-180, 186, 189, 192, 196, 202-203, 208-214

regional, 184

service from, 60, 67, 85

static IP addresses, 101

Connect To dialog box

connections

always-on, 96

breaking, 201

circuit switched, 145, 212

Dial-up, 175, 190

multiple computers, 92

multiple, 17-18, 22, 42, 46, 53, 69, 74, 78, 80, 83, 86, 89, 91-92, 95, 113-116, 123, 130-131, 135, 137, 139, 149, 159, 162, 173, 176, 195, 211

physical, 18-19, 44, 70, 87, 107, 129, 197, 207, 210

Windows 2000, 173

Windows NT, 96

connectivity, 70, 178, 180-181, 183, 197

connectors, 22, 104-105, 123, 131

constant bit rate (CBR), 176, 178

central offices (COs)

connection to, 15, 10, 49, 79, 83, 96, 98, 170, 204

numbers of, 63

cost

consumer electronics bus (CEBus), 161

crosstalk, 18, 24, 27, 31, 33, 103-104

customer premises equipment (CPE)

connections, 2, 6, 12, 14-15, 2, 5, 8-10, 18, 51, 59, 80, 85, 91, 96, 135, 169-170, 173, 175-176, 180, 184, 203

firmware, 85, 123

security, 15, 10-11, 62, 95-97, 99-101, 173, 175, 179, 182, 193

vendors, 195-196, 201

data services, 16, 11-12, 56, 61, 64, 68-69, 84, 117, 134, 144-146

data transmission, 13-16, 9-12, 18-19, 21-22, 24-27, 30, 34, 41-42, 44-47, 49-52, 54-58, 70, 72-73, 75, 79, 91-92, 104-105, 114, 118-119, 121, 128, 139-140, 145, 147-148, 153-156, 158-161, 168-170, 172-173, 176, 189, 202-203, 213

deployment

telecommunications industry, 158

Dial-up connections

Windows 2000, 173

Dial-Up Networking (DUN)

connections, 2, 6, 12, 14-15, 2, 5, 8-10, 18, 51, 59, 80, 85, 91, 96, 135, 169-170, 173, 175-176, 180, 184, 203

installation, 4, 12, 3, 9, 18, 20, 22, 32, 42, 44, 49, 54, 79-80, 87, 96, 102-104, 125, 136-137, 139, 150, 158, 161, 167

Windows NT, 96

digital communications, 17, 42

digital signal processor (DSP), 123, 154-155

digital subscriber line (DSL)

G.lite, 49, 103

history, 40

types, 5-6, 12, 15, 4-5, 8-9, 18, 24, 28-30, 33, 42, 57, 61, 70, 72, 74, 76, 78-80, 84-86, 89-90, 92, 95-96, 99, 102, 105, 111, 114, 116, 131, 135, 145, 147-148, 150, 158, 163, 167-168,

173, 176-177, 197, 203

digital subscriber line access multiplexer (DSLAM), 49, 52, 81, 108, 112-114, 129, 135-137

digital television (DTV), 46, 53, 78

discrete multitone systems (DMT), 24-27, 47-48

distance, 9, 11-12, 14-15, 6, 8, 10-11, 19, 24, 37, 40, 42-43, 45-47, 51, 54, 65, 79, 91-92, 117, 121, 136-137, 142, 145, 149, 157-158, 161, 179-180, 182-186, 209, 211-213

distance learning, 15, 11, 179-180, 182-186

distribution service, 61, 64

domain names
  transferring, 1, 33, 203, 212

downlink, 63, 65

download speed, 57

downloading
  broadband, 55, 161, 179-184, 187, 189-190, 192, 196, 198-199, 202-203, 206, 211, 213-214
  software, 6-7, 5, 29, 43, 85, 87, 95-96, 99, 101, 122-123, 128-129, 137, 150, 155, 165, 167, 170, 172-173, 175-176, 179, 189, 192, 200, 208, 210-212

drivers, 54, 189, 197, 204

E1, 8, 6, 43-45, 95, 139, 163, 173

e-commerce, 188, 192

electromagnetic interference (EMI), 159

electronic mail (email), 101, 172, 179-180, 185, 203, 205, 211-213

email. See electronic mail
  attachments, 212
  broadband, 55, 161, 179-184, 187, 189-190, 192, 196, 198-199, 202-203, 206, 211, 213-214
  host, 95, 98-99
  servers, 95
  Vmail, 172

entertainment, 15, 11, 89, 126, 168, 179, 181-182, 190, 195, 197-199, 201

encryption, 175

equipment, 2, 19-20, 22-23, 25, 39, 41-42, 44, 48-50, 59-61, 65, 70, 72, 75-76, 78-80, 83-86, 101-103, 107-108, 110-111, 114, 116, 118-139, 141, 143, 145-148, 150, 163, 167-168, 177, 187, 191-192, 196, 199, 206, 210

error correction, 17-18, 29

error detection, 27-29

EtherFast routers
  configuration, 87, 92, 95, 98-99, 101-102, 106-107

Ethernet
  bridges, 85, 153
  connections, 2, 6, 12, 14-15, 2, 5, 8-10, 18, 51, 59, 80, 85, 91, 96, 135, 169-170, 173, 175-176, 180, 184, 203
  hubs, 59, 69, 71, 86, 91, 128
  networks, 2, 9, 11-12, 14, 2, 6, 8, 10, 46, 48, 57, 59, 62, 64, 66-67, 69-71, 75, 78-79, 85, 87, 90-91, 93, 95, 99, 106-107, 113-114, 116, 120, 129, 134, 141, 147, 156, 163, 173, 176-177, 180-181, 200, 203, 208, 210-211, 213
  routers, 85, 107, 114, 116, 128, 130, 170, 173, 176
  Windows 2000, 173
  Windows NT, 96

fiber
  cable, 4, 15, 3-4, 11, 18, 33, 36, 38, 54-57, 59, 61-67, 79, 91, 101, 103-105, 110, 114, 118, 120, 127, 144, 146, 158-159, 161, 199, 201, 211
  ring, 2, 27, 81, 170
  optics, 67

fiber distributed data interface (FDDI), 14, 10

filtering
  packets, 11, 7, 116, 145, 176
  routers, 85, 107, 114, 116, 128, 130, 170, 173, 176

firewalls
  networks, 2, 9, 11-12, 14, 2, 6, 8, 10, 46, 48, 57, 59, 62, 64, 66-67, 69-71, 75, 78-79, 85, 87, 90-91, 93, 95, 99, 106-107, 113-114, 116, 120, 129, 134, 141, 147, 156, 163, 173, 176-177,

180-181, 200, 203, 208, 210-211, 213

personal, 60, 70, 73, 84, 86-88, 92, 94, 102-103, 123, 167, 180, 185, 187, 191, 193, 200, 206, 212, 214

security, 15, 10-11, 62, 95-97, 99-101, 173, 175, 179, 182, 193

video conferencing, 146, 168, 170-172, 184, 203, 209-210

firmware, 85, 123

frame, 49, 119, 135-136, 144, 146

frame relay, 119, 144, 146

frequency, 18, 23, 27, 31-32, 36, 49, 51-52, 56-57, 61, 63-64, 69, 72, 77, 89, 91-92, 104, 109-110, 123, 138, 150, 154, 156-162, 164-165

FTTC. See fiber to the curb

FTTH. See fiber to the home

gateway, 12, 15, 8, 10, 59-60, 84, 89-90, 137-138, 141, 149, 159, 161, 165, 167-168

grounding, 158

hackers, 101

hardware, 42, 85, 101, 123, 128-129, 188-189, 192, 212

HDSL

downstream, 25, 46-47, 49-51, 159, 168

forum, 47-48, 51, 53

modulation, 8, 17-18, 24-25, 41, 45, 48, 53, 83, 85, 153, 156

termination unit (HTU), 45

upstream, 25, 46-47, 49-51, 168

HDSL2

downstream, 25, 46-47, 49-51, 159, 168

forum, 47-48, 51, 53

modulation, 8, 17-18, 24-25, 41, 45, 48, 53, 83, 85, 153, 156

upstream, 25, 46-47, 49-51, 168

high definition television (HDTV), 15, 11, 64, 169

Home Phoneline Networking Alliance (HomePNA), 91

host, 95, 98-99

hubs

wireless, 18, 29, 55-56, 61-75, 84, 88, 91,

118, 120, 126, 134, 206

hybrid fiber coax (HFC), 55, 57, 59-60

IDSL

downstream, 25, 46-47, 49-51, 159, 168

forum, 47-48, 51, 53

modulation, 8, 17-18, 24-25, 41, 45, 48, 53, 83, 85, 153, 156

upstream, 25, 46-47, 49-51, 168

incumbent local exchange carrier (ILEC), 141-142, 145, 149

integrated services digital network (ISDN), 16, 12, 17-20, 22, 41-43, 46, 109, 128, 134, 168

Intelligent Network (IN), 1-13, 15-16, 1-12, 17-19, 21-25, 27-30, 33-37, 43-49, 52-53, 55-59, 61-80, 82, 84-92, 95-96, 98, 100-106, 109-111, 114-115, 117-137, 139-145, 147-150, 154, 156, 158-161, 165-170, 173-176, 178-193, 195-203, 205-213

interconnection, 1, 9, 11-12, 1-2, 5-7, 9, 17, 59, 75, 79-80, 83, 107, 119-120, 134-135, 137, 139-141, 147, 163, 213

International Telecommunications Union (ITU), 74, 170

interference, 18, 33-34, 45-46, 71, 79, 103-104, 111, 158-159

Internet, 1, 12-15, 2, 8-11, 30, 46-48, 56-57, 59, 63-64, 66-67, 79, 83-84, 91, 94-101, 107, 113-114, 116, 119, 134-135, 137, 139-141, 144-145, 147, 149, 159, 163, 167, 170, 172-176, 180-181, 183-184, 187, 189-190, 193-201, 203, 205, 207-208, 210-212

internet protocol (IP)

configuration, 87, 92, 95, 98-99, 101-102, 106-107

internet protocol (IP) addresses

advanced, 10-11, 13, 15, 7, 9, 11, 17, 42-43, 53-54, 59, 67, 75, 86, 91, 99, 109, 118, 120-121, 137, 143-144, 149, 153-156, 168, 179, 182, 191, 209, 211

classes, 64, 76, 183, 185

configuration, 87, 92, 95, 98-99, 101-102, 106-107

dynamic, 95, 98-99, 101, 170-172

private, 12, 8, 64, 68, 70, 78, 95, 97-98, 163, 173-174

settings, 46

static, 101

video conferencing, 146, 168, 170-172, 184, 203, 209-210

Internet Service Providers (ISPs)

terms, 122, 201, 208

web site, 82, 98, 199

internet telephony, 149, 211-212

interoperability, 146

layers, 176

leased lines, 45

loading coils, 31, 36-37, 119, 150

local area network (LAN)

configuration, 87, 92, 95, 98-99, 101-102, 106-107

hardware, 42, 85, 101, 123, 128-129, 188-189, 192, 212

modems, 2, 15-16, 2, 11-12, 42, 46-47, 49, 56-57, 83-86, 111-113, 117, 119, 121-123, 125, 127-130, 134-135, 146-149, 200

local exchange carrier (LEC), 11-12, 8-9, 141-142, 150, 163, 168

local loop

access lines, 20-21, 35, 37, 49, 107, 111, 113, 130, 139, 143, 163, 191

bridge taps, 35-36, 46, 119

loading coils, 31, 36-37, 119, 150

reach, 47, 61, 168, 171, 185, 190, 196, 201, 208

terminating, 167

wires, 2-6, 16, 2-5, 12, 17, 22, 27, 32, 35, 41, 45, 52, 56, 71, 84, 103-104, 109, 111, 113, 154, 156, 173

wiring, 4-5, 3-4, 43, 84, 92, 101, 104, 106, 153, 161-162

local multipoint distribution service (LMDS), 15, 11, 61, 64-65

Macs

connections, 2, 6, 12, 14-15, 2, 5, 8-10, 18,

51, 59, 80, 85, 91, 96, 135, 169-170, 173, 175-176, 180, 184, 203

media, 48, 59, 107, 116, 163, 176, 179, 182, 184, 190, 197-199, 201, 203, 205-207, 209

modem

cable modems, 15, 11, 56-57, 127, 146

modulation

amplitude, 24-25, 154-156

frequency, 18, 23, 27, 31-32, 36, 49, 51-52, 56-57, 61, 63-64, 69, 72, 77, 89, 91-92, 104, 109-110, 123, 138, 150, 154, 156-162, 164-165

phase, 24, 156

quadrature, 24-25

quadrature amplitude (QAM), 24-25

motion pictures expert group (MPEG), 169

MPEG compression, 169

multi-channel/multi-point distribution system (MMDS), 15, 11, 61, 63-64

multiple computers, 92

multicast, 66, 69, 140, 146, 168-169

multiplexing, 20, 51, 80, 120

near video on demand (NVOD), 140, 146

NetMeeting video conferencing software

software, 6-7, 5, 29, 43, 85, 87, 95-96, 99, 101, 122-123, 128-129, 137, 150, 155, 165, 167, 170, 172-173, 175-176, 179, 189, 192, 200, 208, 210-212

NetPhone voice communication

software, 6-7, 5, 29, 43, 85, 87, 95-96, 99, 101, 122-123, 128-129, 137, 150, 155, 165, 167, 170, 172-173, 175-176, 179, 189, 192, 200, 208, 210-212

network

data, 1-2, 8, 11-16, 1-2, 6-7, 9-12, 18-19, 21-27, 29-30, 34, 39, 41-59, 61, 64-75, 78-79, 83-88, 90-92, 94-96, 98-100, 104-105, 109, 112-121, 128, 134-140, 144-148, 150, 153-156, 158-163, 165, 167-170, 172-176, 178-180, 186, 189, 192, 196, 202-203, 208-214

management, 116, 179, 182, 192-195, 205, 208

termination, 45, 49, 87, 102, 104, 110-111, 141, 168
Network Address Translation (NAT)
  private IP addresses, 95
  routers, 85, 107, 114, 116, 128, 130, 170, 173, 176
network interface card (NIC)
  configuration, 87, 92, 95, 98-99, 101-102, 106-107
  modems, 2, 15-16, 2, 11-12, 42, 46-47, 49, 56-57, 83-86, 111-113, 117, 119, 121-123, 125, 127-130, 134-135, 146-149, 200
network interface cards (NICs)
  configuration, 87, 92, 95, 98-99, 101-102, 106-107
  drivers, 54, 189, 197, 204
  installation, 4, 12, 3, 9, 18, 20, 22, 32, 42, 44, 49, 54, 79-80, 87, 96, 102-104, 125, 136-137, 139, 150, 158, 161, 167
  multiple, 17-18, 22, 42, 46, 53, 69, 74, 78, 80, 83, 86, 89, 91-92, 95, 113-116, 123, 130-131, 135, 137, 139, 149, 159, 162, 173, 176, 195, 211
  PC Card, 123
networks
  backbone, 59, 69, 107, 112-113, 116, 120, 136, 147, 163, 176
  cabling, 20, 45, 79, 103, 105, 111, 153, 158
  connectors, 22, 104-105, 123, 131
  firewalls, 99-100, 170, 175
hubs
  proxy servers, 95
  routers, 85, 107, 114, 116, 128, 130, 170, 173, 176
  switches, 6-7, 9, 11, 5-7, 80, 114-115, 120, 175
  wireless, 18, 29, 55-56, 61-75, 84, 88, 91, 118, 120, 126, 134, 206
noise, 3, 18, 28, 31, 103
office buildings, 57, 70, 80
online banking, 193-194
Open Systems Interconnection (OSI)

model, 205
operations, administration & maintenance (OA&M) , 139
packet, 11, 7, 19, 70, 90, 114-115, 145-146, 175, 211-212
packet data, 19, 90, 115, 175, 212
packets, 11, 7, 116, 145, 176
packet switching, 212
pairgain, 17
patent royalty, 121, 125, 127, 132
patch, 1-2, 115
pay-per-view, 61-62, 198
PBX. See Private Branch Exchange
Peripheral Component Interconnect (PCI)
  modems, 2, 15-16, 2, 11-12, 42, 46-47, 49, 56-57, 83-86, 111-113, 117, 119, 121-123, 125, 127-130, 134-135, 146-149, 200
  networks, 2, 9, 11-12, 14, 2, 6, 8, 10, 46, 48, 57, 59, 62, 64, 66-67, 69-71, 75, 78-79, 85, 87, 90-91, 93, 95, 99, 106-107, 113-114, 116, 120, 129, 134, 141, 147, 156, 163, 173, 176-177, 180-181, 200, 203, 208, 210-211, 213
  Windows 2000, 173
  Windows NT, 96
phase modulation (PM), 24
phone lines
  cabling, 20, 45, 79, 103, 105, 111, 153, 158
  networking, 84, 89, 91, 96, 161-162, 170, 175, 203
plain old telephone service (POTS), 2-3, 12, 2-3, 8, 21, 35, 42, 46, 49, 53, 67, 89, 102-103, 108-109, 168
point-to-multipoint, 66, 69
point-to-point, 163, 173, 175
ports, 7, 5, 92, 136
post, telephone and telegraph (PTT), 11, 8, 65
POTS. See Plain Old Telephone Service
  wiring, 4-5, 3-4, 43, 84, 92, 101, 104, 106, 153, 161-162
preparation
primary rate interface (PRI), 19
Private Branch Exchange (PBX), 95, 211

protocols, 12, 30, 45, 47, 83, 111, 163, 168, 173, 176

providers, 13, 16, 10, 12, 56, 65, 67, 79, 107, 116, 118-120, 137, 142-145, 148-149, 158, 163, 170, 175-176, 181, 200, 205, 210

proxy servers

  networks, 2, 9, 11-12, 14, 2, 6, 8, 10, 46, 48, 57, 59, 62, 64, 66-67, 69-71, 75, 78-79, 85, 87, 90-91, 93, 95, 99, 106-107, 113-114, 116, 120, 129, 134, 141, 147, 156, 163, 173, 176-177, 180-181, 200, 203, 208, 210-211, 213

  security, 15, 10-11, 62, 95-97, 99-101, 173, 175, 179, 182, 193

  WinGate, 95

PSTN. See Public Switched Telephone Network

quadrature amplitude modulation (QAM), 24-25

quality of service (QoS), 91, 145, 165, 177, 212

RADSL

  downstream, 25, 46-47, 49-51, 159, 168

  forum, 47-48, 51, 53

  modulation, 8, 17-18, 24-25, 41, 45, 48, 53, 83, 85, 153, 156

  upstream, 25, 46-47, 49-51, 168

Rate Adaptive Digital Subscriber Line (RADSL), 42, 49, 165

reliability, 84, 89, 128, 153

Remote Access Service (RAS)

  installation, 4, 12, 3, 9, 18, 20, 22, 32, 42, 44, 49, 54, 79-80, 87, 96, 102-104, 125, 136-137, 139, 150, 158, 161, 167

repeater, 20, 41

research

  business, 9, 11, 6, 8, 49, 51, 54-55, 59, 67, 80, 103, 109, 115, 127, 133, 135, 143-144, 159-160, 180-181, 183, 186-187, 189, 193, 197, 203-205

residential gateway, 89-90

resolution

router, 12, 85, 114-115, 129, 137, 162

routers

  filtering, 85, 123

  filters, 18, 46, 49, 80, 103, 123, 125, 150

  networks, 2, 9, 11-12, 14, 2, 6, 8, 10, 46, 48, 57, 59, 62, 64, 66-67, 69-71, 75, 78-79, 85, 87, 90-91, 93, 95, 99, 106-107, 113-114, 116, 120, 129, 134, 141, 147, 156, 163, 173, 176-177, 180-181, 200, 203, 208, 210-211, 213

  packets, 11, 7, 116, 145, 176

  security, 15, 10-11, 62, 95-97, 99-101, 173, 175, 179, 182, 193

  video conferencing, 146, 168, 170-172, 184, 203, 209-210

sales, 79, 119, 122, 125-127, 133, 143, 148-149, 186, 188, 190-191, 197-198, 202

SDSL

  downstream, 25, 46-47, 49-51, 159, 168

  forum, 47-48, 51, 53

  modulation, 8, 17-18, 24-25, 41, 45, 48, 53, 83, 85, 153, 156

  upstream, 25, 46-47, 49-51, 168

security

  firewalls, 99-100, 170, 175

  hackers, 101

  proxy servers, 95

  routers, 85, 107, 114, 116, 128, 130, 170, 173, 176

  solutions, 18, 175

  viruses, 101

  Windows, 87, 96, 173, 175, 198

servers

  bandwidth, 12, 15-16, 9, 11, 21-24, 29-30, 42-43, 49, 55-56, 64, 70, 75, 78-81, 120-121, 134, 140, 153-154, 156, 158, 169, 178, 180-181, 190-192, 196-197, 201, 203, 208

  e-mail, 101, 172, 179-180, 185, 203, 205, 212-213

  proxy servers, 95

  proxy, 94-95

  speed, 2, 12, 14, 16, 2, 9-10, 12, 19, 21-23, 26-27, 40, 42-46, 50-51, 53-55, 57, 67, 69, 74-75, 77, 79, 104, 120, 131, 134-135, 144-145,

153, 161, 163, 169, 175-177, 179-181, 183-184, 189, 196, 204, 210, 213

service
  routed, 8-9, 6, 108, 112, 141, 165, 167-168
  sharing, 18, 23, 119, 139-140, 165, 184, 203

service providers, 13, 16, 10, 12, 56, 65, 67, 79, 107, 116, 118-120, 137, 142-145, 148-149, 158, 163, 170, 175-176, 181, 205

setup, 6, 10, 5, 7, 19, 30, 98-100, 115-116, 123, 131, 170, 175

shopping, 179-180, 187-190, 192, 201, 205

shopping around
  bandwidth, 12, 15-16, 9, 11, 21-24, 29-30, 42-43, 49, 55-56, 64, 70, 75, 78-81, 120-121, 134, 140, 153-154, 156, 158, 169, 178, 180-181, 190-192, 196-197, 201, 203, 208
  cost, 1, 14, 16, 1, 10, 12, 18, 20, 22, 40, 43-44, 53, 55, 57, 59, 66-67, 75-76, 79-80, 86, 96, 101, 104, 118-146, 148-150, 155, 161, 179, 181, 183, 192, 194-196, 198, 202-203, 205, 207, 210, 212
  speed, 2, 12, 14, 16, 2, 9-10, 12, 19, 21-23, 26-27, 40, 42-46, 50-51, 53-55, 57, 67, 69, 74-75, 77, 79, 104, 120, 131, 134-135, 144-145, 153, 161, 163, 169, 175-177, 179-181, 183-184, 189, 196, 204, 210, 213
  users, 29, 49, 57-59, 62, 65, 69-70, 81, 96, 101, 139, 147, 150, 159, 168, 170-172, 175, 180-181, 190, 193, 196, 200, 212-213

signaling, 10-12, 7, 19-20, 43, 72, 141, 168

signaling system 7 (SS7), 10-11, 7

simple network management protocol (SNMP), 116

software
  video conferencing, 146, 168, 170-172, 184, 203, 209-210

sound, 165, 167, 169-170, 183, 212

speed
  broadband, 55, 161, 179-184, 187, 189-190, 192, 196, 198-199, 202-203, 206, 211, 213-214
  file transfers, 181

preparation, 59
  servers, 95

splitter
  active, 22, 80, 89, 111, 181, 185, 187
  passive, 80, 89, 111

standards
  competition, 15, 11, 59, 65, 118, 120, 127, 146-147, 158

startup
  charges, 140, 142

stateful packet inspection
  firewalls, 99-100, 170, 175

static IP addresses
  configuration, 87, 92, 95, 98-99, 101-102, 106-107

subscribers
  defined, 195, 200

switched services, 91

switch
  crossbar, 6-7, 5
  time slot interchange, 7-8, 5-6

switches
  networks, 2, 9, 11-12, 14, 2, 6, 8, 10, 46, 48, 57, 59, 62, 64, 66-67, 69-71, 75, 78-79, 85, 87, 90-91, 93, 95, 99, 106-107, 113-114, 116, 120, 129, 134, 141, 147, 156, 163, 173, 176-177, 180-181, 200, 203, 208, 210-211, 213

synchronous digital hierarchy (SDH), 51

synchronous optical network (SONET), 51

Synchronous Digital Subscriber Line (SDSL), 41-42, 45, 137-138, 166, 211

T1, 8, 6, 20, 22, 40, 43-45, 95, 119, 121, 139-140, 144, 163, 173

Telecommunications Act of 1996, 65

telecommuting, 203

telemedicine, 15, 11, 179, 182, 208-210, 214

time delay, 155

time slot interchange (TSI), 7-8, 5-6, 129

trunk, 4, 3, 17, 59

unspecified bit rate (UBR), 176, 178

variable bit rate (VBR), 176

VDSL2

downstream, 25, 46-47, 49-51, 159, 168

forum, 47-48, 51, 53

modulation, 8, 17-18, 24-25, 41, 45, 48, 53, 83, 85, 153, 156

upstream, 25, 46-47, 49-51, 168

Very High Speed Digital Subscriber Line (VDSL), 16, 12, 42-43, 48, 50-53, 81-82, 109, 111, 165, 168, 211

video on demand (VOD), 11, 61, 80-81, 140, 146, 168-169

voice over internet protocol (VoIP), 167

wideband, 55, 68, 71